W0080036

# Cardiovascular, Respiratory, Gastrointestinal and Genitourinary Malformations

**Advances in the Study of Birth Defects**

VOLUME 6

# Cardiovascular, Respiratory, Gastrointestinal and Genitourinary Malformations

EDITED BY

## T. V. N. Persaud

and

## M. P. Persaud

MTP PRESS LIMITED
*International Medical Publishers*

Published by
MTP Press Limited
Falcon House
Lancaster, England

Copyright © 1982 MTP Press Limited

Softcover reprint of the hardcover 1st edition 1982
First published 1982

All rights reserved. No part of this publication
may be reproduced, stored in a retrieval
system, or transmitted in any form or by any
means, electronic, mechanical, photocopying,
recording or otherwise, without prior permission
from the publishers.

**British Library Cataloguing in Publication Data**
Cardiovascular, respiratory, gastrointestinal
    and genitourinary malformations.—(Advances in the
    study of birth defects; v. 6)
    1. Abnormalities, Human
    I. Persaud, T.V.N.          II. Series
    616'.043          RG626
ISBN-13: 978-94-011-7958-4          e-ISBN-13: 978-94-011-7956-0
DOI: 10.1007/978-94-011-7956-0

Typeset by Macmillan India Ltd., Bangalore

# Contents

v

# List of contributors

**J. C. BEAR**
Faculty of Medicine
Memorial University
St John's, Newfoundland
Canada A1B 3V6

**S. CYWES**
Department of Paediatric Surgery
University of Cape Town and
Red Cross War Memorial Children's
Hospital, Rondebosch 7700, Cape Town,
South Africa

**M. R. Q. DAVIES**
Department of Paediatric Surgery
University of Cape Town and
Red Cross War Memorial Children's
Hospital, Rondebosch 7700, Cape Town,
South Africa

**L. G. DIXON**
Department of Pathology
Children's Hospital of Los Angeles
4650 Sunset Boulevard
Los Angeles
California 90027
USA

**M. GHARIB**
Kinderchirurgische Klinik des
Städt.-Kinderkrankenhauses Köln-Riehl
Amsterdamer Str. 59
5000 Köln 60
German Federal Republic

**R. E. HARRIS**
Suite 171
Methodist Plaza
4499 Medical Drive
San Antonio, Texas 78238, USA

**P. K. KOTTMEIER**
Pediatric Surgery
State University of New York
Downstate Medical Center
450 Clarkson Avenue
Brooklyn, NY 11203, USA

**B. H. LANDING**
Departments of Pathology
and Pediatrics,
Children's Hospital of Los
Angeles and University of
Southern California School
of Medicine,
4650 Sunset Boulevard
Los Angeles,
California 90027
U.S.A

**R. I. MACPHERSON**
Department of Radiology
Richland Memorial Hospital
3301 Harden Street
Columbia
South Carolina 29203
USA

**J. A. NOORDIJK**
Department of Paediatric Surgery
Sophia Children's Hospital
Gordelweg 160
3038 GE Rotterdam
The Netherlands

**M. L. RAMENOFSKY**
Division of Pediatric Surgery
Tufts University School of Medicine
Boston Floating Hospital
171 Harrison Avenue
Boston, Mass. 02111, USA

**H. RODE**
Department of Paediatric Surgery
University of Cape Town and
Red Cross War Memorial Children's
Hospital, Rondebosch 7700, Cape Town,
South Africa

**T. R. WELLS**
Department of Pathology
Children's Hospital of Los Angeles
4650 Sunset Boulevard,
Los Angeles, California 90027, USA

**C. A. SALVATORE**
Gynecologic and Obstetric Department
The University of São Paulo
Medical School
São Paulo, Brazil

**N. E. WISEMAN**
Department of Surgery
Winnipeg Children's Hospital
685 Bannatyne Avenue
Winnipeg, Manitoba, Canada R3E 0W1

**A. A. SHEM-TOV**
Heart Institute
The Chaim Sheba Medical Centre
Tel-Hashomer
Israel

# Preface

Birth defects have assumed an importance even greater now than in the past because infant mortality rates attributed to congenital anomalies have declined far less than those for other causes of death, such as infectious and nutritional diseases. As many as 50% of all pregnancies terminate as miscarriages, and in the majority of cases this is the result of faulty intrauterine development. Major congenital malformations are present in at least 2% of all liveborn infants, and 22% of all stillbirths and infant deaths are associated with severe congenital anomalies. Not surprisingly, there has been a great proliferation of research into the problems of developmental abnormalities over the past few decades.

This series, *Advances in the Study of Birth Defects*, was conceived in order to provide a comprehensive focal source of up-to-date information for physicians concerned with the health of the unborn child and for research workers in the fields of fetal medicine and birth defects. The first four volumes featured recent experimental work on selected areas of high priority and intensive investigation, including mechanisms of teratogenesis, teratological evaluation, molecular and cellular aspects of abnormal development, and neural and behavioural teratology. It seems logical and timely that the clinical aspects should now be presented. Accordingly, leading experts were invited to review a broad range of common problems from the standpoint of embryology, aetiology, clinical manifestations, diagnosis and management. This volume deals with cardiovascular, respiratory, gastrointestinal, and genitourinary malformations.

I am greatly indebted to the distinguished panel of contributors. Their enthusiasm, many useful suggestions and cooperation have made this volume a reality. As co-editor of this volume, Dr M. P. Persaud has provided valuable support and advice. My sincere thanks go to the publishers, especially Mr D. G. T. Bloomer, Managing Director, MTP Press Limited, for their encouragement and for extending to me every kindness. Once again, I owe much to my secretary, Mrs Barbara Clune, for her invaluable and unstinting help. I should also like to thank Mr Roy Simpson, medical photographer, for his assistance with several of the illustrations.

This work is affectionately dedicated to Indrani, Sunita and Ren.

Winnipeg, Canada                                   T. V. N. Persaud<br>January, 1981

# 1
# Diagnosis of cardiovascular malformations

A. A. SHEM-TOV

As we are all aware, there is an evolution in the concept of diagnosis and management of patients with congenital heart disease. Early palliative surgical procedures were followed by reparative surgery and are now to a large extent replaced by the latter.

This new approach to the problem has been made possible by the collaboration of a team of cardiologists, surgeons, anaesthetists and nurses to the extent that reparative surgery can be accomplished in the first weeks or months of life.

The problems present in the neonatal period are, for the most part, emergency problems where precise and early diagnosis are essential elements of success.

The disorders which patients in the neonatal period may present include cyanosis, congestive heart failure, arrhythmias and murmurs.

Cyanosis is a serious symptom in newborn infants. Once it can be established that this serious symptom does not have a pulmonary origin produced by cerebral or thoracic causes, or peripheral origin from sepsis, or from cold or over-transfusion, a cardiovascular aetiology should be looked for. Persistent cyanosis, especially during the first day or two of life, is an ominous sign. There is no time to wait for possible improvement, since the situation can only become worse. Early diagnostic procedures have real advantages before severe acid base disturbance has developed; thus the infant will tolerate the diagnostic procedures well and also will be in the best possible condition for any subsequent surgical intervention.

Needless to say, for babies with complete transposition of the great arteries, balloon septostomy, which can be performed during catheterization, is a lifesaving procedure. Thanks to this procedure, babies with complete transposition of the great arteries are followed in the outpatient clinics – something not frequently seen before.

Babies with complete transposition of the great arteries present with early cyanosis, with or without congestive heart failure, with a more or less typical radiological picture, the electrocardiogram showing right ventricular hyper-

1

trophy or biventricular hypertrophy. Occasionally the ECG may be normal and this finding is helpful in diagnosing this disease, since this is the only cyanotic heart disease where the electrocardiogram could be normal. Murmurs may or may not be present with the second sound being mostly single. All these signs and symptoms facilitate the diagnosis, and immediate catheterization can be performed. It should be stressed that in complete transposition of the great arteries cyanosis can be present without congestive heart failure, but congestive heart failure cannot be present without cyanosis.

The problem in the usual tetralogy of Fallot in the newborn may not be acute. Cyanosis may be very mild and because of the progressive nature of right ventricular outflow obstruction, a significant left to right shunt may be present early in life in some cases with congestive heart failure. A splitting 2nd heart sound may also be present. The radiological picture of uplifted apex, absent main pulmonary artery and especially right aortic arch, when present, will render the clinical diagnosis possible. In severe tetralogy of Fallot, especially with pulmonary atresia, the clinical picture is different. The cyanosis is severe, the 2nd sound is single, and a weak continuous murmur of patent ductus arteriosis is frequently present. Emergency catheterization is mandatory to visualize the anatomy, especially of the pulmonary arteries, so that one of the shunting procedures can be performed when necessary. In rare cases, the administration of prostaglandin E may be necessary to keep the ductus patent until a shunting operation can be performed.

Severe pulmonary stenosis is another problem in the neonatal period. The baby is blue with right ventricular failure. The physical examination may be diagnostic with a systolic click and murmur, and a split 2nd sound with weak pulmonic component. A pansystolic murmur of tricuspid insufficiency is generally present. Cardiomegaly with ischaemic lung is seen on the X-ray picture, and the ECG shows not only right ventricular hypertrophy but also significant left ventricular potentials. An emergency catheterization is carried out followed immediately by pulmonary valvotomy. The relatively stable condition in some babies may be misleading, and postponing operation may endanger the baby's life. In some cases the pulmonary valve may be dysplastic. Exact preoperative diagnosis should be made because the surgical approach is different in this condition.

Total anomalous pulmonary venous connection, especially of the supra-diaphragmatic type, may not present as a serious problem in the neonatal period; cyanosis is mild; congestive heart failure may appear late, but the clinical picture of an atrial septal defect (parasternal heave, ejection systolic murmur, almost fixed splitting of the 2nd sound, mid-diastolic flow murmur) together with cyanosis, should render the diagnosis possible. The presence of extreme right ventricular hypertrophy in the electrocardiogram with Q-wave in $V_1$ and absence of left ventricular potential help greatly in the diagnosis.

These are some of the cyanotic heart malformations which may present in the newborn as emergency problems. Early diagnosis is necessary in view of surgical intervention when needed, or for an accurate diagnosis and outlying plan of action for the near future.

Congestive heart failure may be present in cyanotic heart disease, but, on the other hand, it may be the presenting problem in babies with cardiac disorders.

Early signs of heart failure should be looked for before the neonate enters into a fullblown picture of decompensation. Increased respiratory rate, tachycardia or gallop, loud heart sounds, liver enlargement, questionable cyanosis are all early warnings which demand prompt administration of anticongestive treatment. The administration of this treatment should not lower the vigilance of the paediatrician – close hourly follow-ups should be enforced with a view to emergency studies if necessary. Like cyanosis in cyanotic heart disease, congestive heart failure of cardiovascular origin in the neonate is a serious problem and demands a precise and early diagnosis.

The hypoplastic left heart syndrome should be recognized promptly for obvious reasons. Early severe heart failure in a pale, questionably cyanotic baby, pulseless, with enlarged heart and severe pulmonary congestion and an electrocardiogram showing extreme right ventricular hypertrophy with Q-wave in $V_1$ should raise the possibility of this diagnosis. The echocardiographic findings of a small left ventricle and aorta should confirm the clinical diagnosis and catheterization can be avoided.

Symptomatic valvular aortic stenosis, although rare, is another emergency problem in the early period of infancy. The physical findings of a systolic click and murmur, single or narrowly split 2nd sound and left ventricular hypertrophy on the electrocardiogram may be present and helpful for the diagnosis. Catheterization should be carried out urgently to confirm the severity of the obstruction. When this is severe, surgery should be performed before one is faced with acute left ventricular failure or terminal acute heart dilatation and shock. Coarctation of the aorta alone rarely presents a serious problem in the newborn. Symptomatic infants with clinically diagnosed coarctation should be catheterized to reveal possible associated malformations such as a ventricular septal defect or a patent ductus arteriosis which are, for the most part, responsible for the heart failure. Because of the constant presence of pulmonary hypertension in these symptomatic infants, right ventricular hypertrophy in the electrocardiogram is the rule. Surgery, when necessary, should be aimed at correcting the associated malformation.

Ventricular septal defect with congestive heart failure is rarely encountered in the neonate because of delayed maturation of the pulmonary vascular bed. Symptomatic ventricular septal defect occurs after the age of 1 month. The clinical picture is that of congestive heart failure. The physical findings are those of a harsh systolic murmur – narrowly split 2nd sound with accentuated pulmonic component, and mid-diastolic flow murmur at the apex. The electrocardiogram shows biventricular hypertrophy and an enlarged heart, and increased pulmonary vascularity is seen on the chest X-ray. Cardiac catheterization should be performed to confirm the diagnosis, and to rule out the presence of patent ductus arteriosis, which if present and large should be ligated.

Arteriovenous (a-v) fistula is rare but can present as a serious problem in the neonate. Clinically, the baby with a large a-v fistula presents with early congestive heart failure, cardiomegaly and pulmonary congestion on the chest X-ray and biventricular hypertrophy on the electrocardiogram. The continu-

3

ous bruit confirms the diagnosis. In order to rule out this possibility one must routinely auscultate the skull, neck, thorax, liver and kidney areas in all newborns.

In our experience with neonates, we have encountered a clinical picture which we should like to call 'transitory congestive cardiomyopathy'. These babies present with congestive heart failure, extremely pathological electrocardiogram and cardiomegaly. They respond well to medical treatment and all symptoms and signs regress and return to normal within a few months. Early in our experience, before we learned to recognize this clinical picture, we catheterized two of these babies and no unusual findings were present.

Endocardial fibroelastosis is not usually a problem of the neonate – the baby with this disease goes into failure after the age of 1 month. Beside congestive heart failure, the electrocardiogram shows left ventricular hypertrophy and strain. The chest X-ray shows cardiomegaly. Catheterization should be performed to rule out an anomalous origin left coronary artery from the pulmonary artery. Digitalis treatment is given and should be continued for as long as symptoms and signs exist.

Patent ductus arteriosis in the neonate presents a great challenge both to the neonatologist and the paediatric cardiologist. When patent ductus arteriosis is diagnosed in the premature baby, active and hourly follow-ups are obligatory in view of the great risk involved. It is obvious that the best conditions should be realized to pull the premature baby in distress out of the critical condition. In some premature babies, particularly in the wake of a respiratory distress syndrome, patent ductus arteriosis has to be viewed with extreme gravity. Medical treatment should be tried first and if not successful, and the baby's condition allows it, indomethacin may be tried. If both these attempts fail, surgical closure of the ductus may be seriously contemplated.

Arrhythmias, and in particular episodes of supra-ventricular tachycardia, in the modern newborn are not rare conditions. The baby is restless and heart failure may occur if the attack lasts for hours. Usually there is no underlying heart condition. After the termination of the attack, Wolff–Parkinson–White Syndrome should be looked for. It is generally agreed that those attacks which appear early in the first weeks of life have a better prognosis. Intravenous Verapamil is recommended to terminate the attack, and with frequent success. The use of digitalis should be avoided because electrical shock is sometimes needed to terminate resistant tachycardia.

Congenital complete heart block should be recognized during gestation. When alone, complete atrioventricular block is asymptomatic. When symptoms are present, they are usually due to associated cardiac malformations such as patent ductus arteriosis, coarctation, endocardial fibroelastosis or corrected transposition of the great arteries. Patent ductus arteriosis should be ligated and coarctation resected to relieve symptoms. When endocardial fibroelastosis is present, implantation of a permanent pacemaker may be indicated to stop evolution and hopefully 'cure', at least clinically, the endocardial fibroelastosis.

Having discussed some of the problems of cardiovascular origin of the newborn, a brief mention should be made of some of the pertinent physical findings, palpations and observations of the neonate which help in the clinical diagnosis. Needless to say, this is not a futile intellectual exercise. Reducing to a minimum the differential diagnosis, helping to decide on the indication of diagnostic approach and procedures and outlining the plan of action in individual patients are vital measures.

(1) Quiet heart in a non-cyanotic baby indicates either a normal heart or a heart disease without critical conditions at the time of examination.

(2) In a non-cyanotic or questionably cyanotic baby with tachypnoea, dyspnoea, tachycardia, but with a quiet heart on palpation, the problem is probably pulmonary and not cardiac.

(3) An active heart in the first place indicates heart disease, and in a non-cyanotic baby possibly means a shunting defect.

(4) In cyanotic heart disease with a mildly active heart one is probably dealing with an obstruction to pulmonary blood flow. On the other hand, an active heart accompanied by cyanosis probably means mixing blood malformation and transposition complex in particular.

(5) Single 2nd sound may be due to aortic closure as in pulmonary atresia or truncus. But when it is due to pulmonic closure, hypoplastic left heart syndrome with aortic atresia may be the cause.

(6) The 2nd heart sound plays an important role in the clinical diagnosis. Split 2nd sound means that both semilunar valves are present, thus ruling out pulmonary atresia and almost always truncus arteriosis.

(7) Split 2nd sound with accentuated pulmonic component means pulmonary hypertension. A weak pulmonic component means pulmonary stenosis.

(8) Fixed splitting of the 2nd sound indicates atrial septal defect.

(9) Paradoxical splitting of the 2nd heart sound means severe obstruction to left ventricular egress or complete left bundle branch block.

(10) So-called 'pink face' malformation with cyanosis of the lower part of the body means interruption of the aortic arch. On the other hand, differential diagnosis with cyanosis of the upper part of the body means interruption of the aortic arch with complete transposition of the great arteries.

These are only some of the physical observations which greatly aid in the clinical diagnosis, and their importance cannot be overemphasized. These clinical observations together with the electrocardiogram and chest X-ray picture, in experienced hands, can be most helpful tools in the diagnosis of cardiovascular malformations.

# 2
# Pathogenetic considerations of respiratory tract malformations in humans

B. H. LANDING, WITH THE TECHNICAL
ASSISTANCE OF T. R. WELLS AND L. G. DIXON

## GENERAL ASPECTS OF THE DEVELOPMENT OF THE HUMAN RESPIRATORY TRACT

The basic timetable of the embryonic human respiratory tract is generally accepted, although disagreement exists on various points. It can be summarized as follows[1,2]:

| embryonic age | anatomical feature |
|---|---|
| 24 d | tracheal bud forms |
| 26–28 d | stem bronchial buds form (but see below) |
| 35 d | lobar bronchi form |
| 5–12 weeks | further bronchial branchings occur |
| 12–25/26 weeks | tracheal glands develop cranio-caudally |
| 16 weeks | terminal bronchioles form = glandular stage of lung |
| 16–24 weeks | respiratory bronchioles form (three orders by birth) = canalicular stage of lung |
| 24–26 weeks | alveoli begin to form = alveolar stage of lung. |

### Growth of peripheral respiratory tract

Although, as is stated above, the central bronchial tree is fully formed by gestational age 16 weeks, some workers conclude that peripheral branching orders continue to be formed throughout intrauterine life and into the postnatal period, although opinion is not totally unanimous[3]. Thus, at birth there are three orders of respiratory bronchioles and one order of alveolar ducts, whereas by age 2 months the adult pattern, with four orders of respiratory bronchioles and three orders of alveolar ducts, is established.

7

Since the number of branchings in the airway to the level of the terminal bronchioles differs significantly in different radii of the lungs[4] – for example, counting from the segmental bronchi the number of such branchings in the lingular, the right middle lobe and the posterior basal segments of the right lower lobe is up to 25 but in the apical segments of the upper lobes only to 20 – the presumption of contemporaneity of same-order branchings further suggests that additional branchings appear after birth. Horsfield and Cumming[5] record even greater range of branching numbers in different areas of the lung, with from eight to 25 orders down to the level of lobular bronchi (reading from the alveoli, the first branches with diameter 0.07 cm (700 $\mu$m) or greater), and two to seven intralobular bronchiolar orders, for a range of ten to 32 branchings from the carina through the respiratory bronchioles. Important to such considerations is recognition of the fact that the segmental bronchi do not all represent the same branching order, nor do many corresponding segmental bronchi belong to the same order in both lungs, thus[7]:

| segmental bronchi, standard nomenclature | | generation |
|---|---|---|
| right lung | 1, 2,3 | 4 |
| | 4, 5, 6 | 5 |
| | 7, 8 | 6 |
| | 9, 10 | 7 |
| left lung | 1, 2 | 6 |
| | 3, 4, 5 | 5 |
| | 6 | 4 |
| | 8 | 5 |
| | 9, 10 | 6. |

The formation of new branchings in the peripheral respiratory tract postnatally accounts for the fact that, although the number of cartilage-bearing bronchi increases during the first postnatal months, the number of cartilage-free peripheral bronchi is, in different radii, independent of the total length of the pathway – namely, is a constant number from the periphery[7]. Matsuba and Thurlbeck[6,7] demonstrated that, although the number of small airways (nonrespiratory bronchioles less than 2 mm in diameter), was negatively correlated with ( = was constant despite difference in) body length, the total number of alveoli was positively correlated with body length (namely, the number of alveoli per such small airway is greater in taller persons).

Despite these considerations, which make it difficult to assign precise single values to many respiratory tract dimensions, it is informative to compute 'increase ratios' for various aspects of the lungs, as is illustrated in Table 2.1, which is based on original data from Weibel[8] and hence subject to modification depending on the 'normal' values one prefers. (Note also that Weibel[9] has emphasized that use of progressively high magnifications has significantly increased the value determined for alveolar surface area, by the 'coast of Britain' effect.)

**Table 2.1**

| | No. of alveoli ( $\times 10^6$ ) | Increase ratio | Air/tissue interface (m²) | Increase ratio | Body surface area (m²) | Increase ratio | LSA/BSA |
|---|---|---|---|---|---|---|---|
| Birth | 24 | 1 | 2.8 | 1 | 0.21 | 1 | 13 |
| 3 months | 77 | 3 | 7.2 | 2.5 | 0.29 | 1.5 | 24 |
| 7 months | 112 | 5 | 8.4 | 3 | 0.38 | 2 | 22 |
| 13 months | 129 | 5 | 12.2 | 4 | 0.45 | 2 | 27 |
| 4 y | 257 | 10 | 22.2 | 8 | 0.67 | 3 | 33 |
| 8 y | 280 | 12 | 32.0 | 11 | 0.92 | 5 | 34 |
| Adult | 296 | 12 | 75.0 | 27 | 1.90 | 10 | 39 |

'Increase ratios' for alveolar number and alveolar surface area with age through childhood showing that adult alveolar number is reached by approximately 8 years, but that over half of alveolar enlargement occurs after this age. The ratio, lung surface area/body surface area, which approaches adult values by age 4 years, demonstrates the significantly smaller margin of lung absorptive capacity versus metabolic oxygen demand in the neonate, the ratio doubling over the first 3 postnatal months

*Abnormality of airway branching numbers*

Abnormal branching patterns are seen in the pulmonary isomerism syndromes, considered below, and in other selective segmental bronchial deficiencies, but general derangement of branching numbers also occurs. Reid[2] discusses a number of such situations, including:

(1) A patient with unilateral agenesis of the lung, whose remaining lung had less than half the normal number of airway generations but twice the normal alveolar number, showing that alveolar multiplication and differentiation do not require a normal number of peripheral airways (the same basic point is made by the normal postnatal increase in alveolar number shown in Table 2.1).

(2) Congenital diaphragmatic hernia, with reduction of airway number and alveolar number in both lungs, more markedly so in the more hypoplastic lung ipsilateral to the hernia.

(3) Erythroblastosis fetalis, with arrest of airway multiplication at gestational age 8–10 weeks, and hence not due to pulmonary compression from diaphragmatic elevation due to hepatosplenomegaly.

(4) Pulmonary hypoplasia associated with renal anomalies (Potter syndrome, oligohydramnios tetrad), with reduction in airway and alveolar number reflecting arrest of development at an early fetal age, before the development of oligohydramnios.

(5) Lobar emphysema due to polyalveolar lobe (airway branching normal, but alveolar number increased fivefold).

(6) Hypoplastic emphysema, a lesion with both reduced airway branching and alveolar number, but with the alveoli markedly enlarged.

## Structural features of the airway

### Muscle

Matsuba and Thurlbeck[6] analysed the composition of small airways (non-alveolated bronchioles of less than 2 mm internal diameter) compared to that of major bronchi. Their data, given in Table 2.2, illustrate the significantly

Table 2.2  Data on composition of airways given by Matsuda and Thurlbeck[6]

|  | Major bronchi | | | | | |
|---|---|---|---|---|---|---|
|  | % Muscle | % Cartilage | % Connective tissue | % Gland | Small airways % Muscle | No. of patients studied |
| Normal children | 2.8 | 38.4 | 42.2 | 17.1 | 9.8 | 11 |
| Children with bronchiolitis | 3.3 | 33.6 | 39.7 | 23.4 | 9.5 | 3 |
| Children with cystic fibrosis of pancreas | 3.7 | 35.7 | 40.8 | 20.1 | 9.0 | 3 |
| Adults | 2.7 | 44.2 | 40.1 | 11.0 | 14.2 | 5 |

higher proportion of muscle (3–4 times) in small versus major air passages, obviously explaining the effect of smooth muscle contraction on this airway level (in addition to the much greater effect on the cross-sectional area of the same decrease in radius for small versus larger passages). To our knowledge, no definite demonstration of a disease with primary quantitative abnormality of small airway muscle has been published. Wailoo and Emery[10] have studied the structure of the pars membranacea of the trachea in detail. The transverse muscle bundles, which lie predominantly superficial to the bodies of the respiratory mucoserous glands, are uniformly arranged in a ladder-like lattice, so that they do not connect one cartilage ring to another. These transverse muscles connect to the inner and medial surfaces of the rings a short distance from their tips. The longitudinal muscle, which lies generally deep to the respiratory glands, on the other hand, is present only in the lower half of the trachea in 32% of children, is present only in the lower third in 12%, and is absent in 16%. Since, in their data, 31% of fetuses of 24 weeks' gestation had longitudinal muscle in the midtrachea, whereas over 50% of children from term on had longitudinal muscle at this level, their 28% of cases with absence or hypoplasia (presence only in the lower third of the trachea) of the longitudinal muscle may be illustrating that an influence operating in late pregnancy can arrest the development of the longitudinal muscle. Study of other aspects of patients with this finding, and of the properties of the later stages of the pregnancies which produced them, would be of interest, as would pathogenetic information on the lesion described as focal muscular hyperplasia of the trachea[11].

### Respiratory glands

The distribution and development of the tracheobronchial mucoserous glands have been described in detail in a series of papers by Tos[12-15]. The distribution of the gland orifices is basically regular, but, since they tend to lie along folds, the orifices tend to occur in longitudinal rows, especially in the lower trachea. The gland mass is predominantly in the intercartilage spaces, with two to six rows of glands per space, the basic pattern appearing to be a symmetrical five-row pattern, with (1) a central row of glands with short straight vertical ducts, (2) two adjoining rows of glands with longer oblique ducts slanting toward the cartilages and (3) two rows of glands lying along the edges and on the surfaces of the rings, with long oblique ducts opening above the luminal faces of the cartilages. The number of glands is fully formed by gestational age 25–26 weeks, so that, as with eccrine sweat glands, the total number does not change after birth. The numbers of glands in the trachea given by Tos[13,14] are shown in Table 2.3. Except that the respiratory glands, like the sweat glands, are essentially absent in anhidrotic ectodermal dysplasia, we know of no reports of abnormality of their number or pattern of distribution in specific diseases, or in regular association with anomalies of the respiratory tract.

### Respiratory tract cartilages
Discrepant data exist on the schedule of development of the respiratory tract cartilages, which form first in the proximal trachea, and progressively appear

**Table 2.3**

|                                  | Children              | Adults                              |
| -------------------------------- | --------------------- | ----------------------------------- |
| Glands in pars membranacea       | 792                   | 818                                 |
| Glands in cartilaginous portion  | 3113                  | 3123                                |
| Total glands in trachea          | 3905                  | 3941                                |
| Density of glands   birth        | $7-10/mm^2$           | $0.7-0.9/mm^2$ $(= 1 - 1.5\,mm^2/gland)$ |
| 8 y                              | $2.2/mm^2$            |                                     |

Data of Tos[14,15] on numbers and distributions of mucoserous respiratory glands in the tracheal wall, illustrating the lack of increase in glands after birth (actually, after 25 weeks of fetal life), and their spreading apart as the airway grows

peripherally; published data on the time of first appearance of cartilages vary from 4 weeks[2] and 7 weeks[4] to 10 weeks[16]. The cartilages of the central passages are stated to be formed by 25 weeks, with more peripheral cartilage plates continuing to form through 48 weeks (2 months postnatally)[16]. In the course of studies on growth abnormalities of the right main bronchus we have assembled data which show an increase in the number of cartilage rings in both the right and left main bronchi after 25 weeks fetal age, as well as during the first year of postnatal life (Table 2.4) (Figure 2.1), as determined by counting the rings visible in dissected specimens stained with toluidine blue at pH 2 and then cleared[17] (see Figure 2.2, below). Since there is no reason to believe that new rings actually form in the central respiratory tree at this late embryonic stage, the apparent explanation of the increased ring count with increasing developmental age is either or both of proximal extension of the carina or shift of the orifice of the right upper lobe bronchus by narrowing of

**Table 2.4**

| | Number specimens studied | Mean no. cartilage rings, RMB | Mean no. cartilage rings/ LMB | Mean ratio LMB/ RMB | Mean length RMB (cm) | Mean length LMB (cm) | Mean ratio LMB/ RMB | Mean no. cartilage rings/cm RMB | Mean no. cartilage rings/cm LMB | Mean ratio RMB/ LMB |
|---|---|---|---|---|---|---|---|---|---|---|
| Premature infants | 20 | 2.95 | 9.05 | 3.07 | 0.33 | 1.13 | 3.42 | 8.9 | 8.0 | 1.11 |
| Male | 12 | | | | | | 3.27 | | | |
| Female | 8 | | | | | | 3.64 | | | |
| Infants < 1 y, not premature | 26 | 3.42 | 9.24 | 2.70 | 0.45 | 1.39 | 3.09 | 7.6 | 6.6 | 1.15 |
| Children > 1 y | 6 | 4.66 | 10.50 | 2.25 | | | 2.82 | | | |

Data demonstrating that the length and the number of cartilage rings in both right and left main bronchi increase during infancy, and that the number of rings per cm of length falls (more rapidly for the LMB than for the RMB)

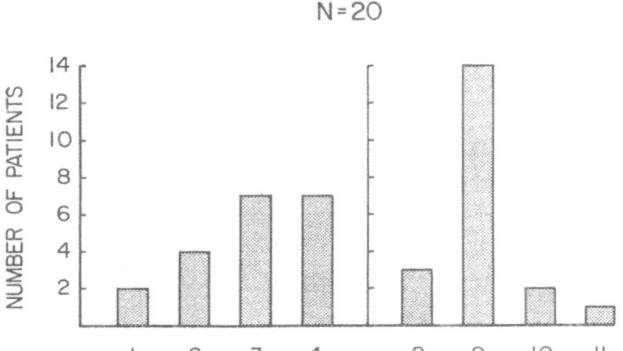

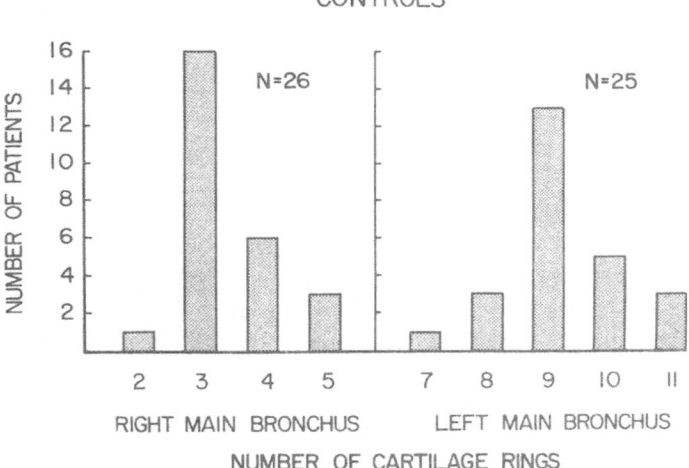

**Figure 2.1** Numbers of cartilage rings in right and left main bronchi of premature infants and of non-premature infants less than 1 year of age (controls). Right main bronchi with two rings or fewer, and left main bronchi with seven rings or fewer, are statistically abnormal

its superior aspect (in effect, by converting a funnel-shaped to a tubular proximal bronchus).

The tracheobronchial cartilage rings begin to form anteriorly on the passages and grow laterally[18]; this presumably explains the fact that the transverse muscle bundles insert on the cartilages lateral to their postero-medial ends – namely, the cartilages continue to grow after the muscle bundles form and attach.

Wailoo and Emery[19] have also studied the structure of the trachea in children with oesophageal atresia (EA) and tracheo-oesophageal fistula

(TEF). In 30 of 40 (75%) specimens, the cartilages were reduced in length. In 6 (15%) the ring-shortening involved the region of the fistula, in 15 (38%) it involved approximately half the trachea, and in 9 (23%) it involved the whole trachea; in only 4 (10%) of their specimens was the length of the cartilage rings of the trachea normal at all levels. As would be expected, the transverse muscle bundles were increased in length in 24 specimens (60%), in 22 (55%) in the region of the fistula and in 2 (5%) over the length of the trachea. These workers found the normal ratio of cartilage to muscle in the infant trachea to be 4.5/1; in patients with EA–TEF the ratio was reduced in 27 (67%) and normal in 13 (33%). The short cartilage rings of the tracheas of patients with EA–TEF are not an indication of tracheal hypoplasia, since in 26 (65%) of Wailoo and Emery's patients the internal perimeter of the trachea was larger than normal. In 19 (48%) this tracheal dilatation was seen only at the level of the fistula, whereas in 7 (18%) it was diffuse; in 9 (23%) of their patients there was reduced internal tracheal perimeter, and in only 5 (13%) was it normal. Wailoo and Emery propose that the wide flat tracheas, with relative cartilage deficiency, of infants with EA–TEF are 'floppy' and dispose these infants to respiratory distress. Our own studies confirm the anatomical findings of Wailoo and Emery. The cartilage deficiency of the tracheas of a high proportion of infants with EA–TEF may explain the high incidence of tracheal compression by abnormal aortic arch vessels seen in this syndrome, as is discussed below. Abnormally wide flat trachea, with wide pars membranacea, was described in 29% of a series of older adult males by Liddelow and Campbell[20], who suggested that the presumed greater compressibility of the trachea with this shape might dispose such persons to some forms of respiratory tract disease. Mackenzie et al.[21] analysed the cross-sectional shapes of adult human tracheas, finding six different shapes.

(1) C-shaped (circular except for the pars membranacea) – 49%.
(2) U-shaped (larger A–P than transverse diameter) – 27%.
(3) D-shaped (larger transverse than A–P diameter, the pattern mentioned above) – 13%.
(4) Elliptical (larger transverse than A–P diameter, but with narrow pars membranacea) – 8%.
(5) True circular – 2%.
(6) Triangular (with an anterior 'keel') – 2%.

With the reservation that comparable studies have not been performed on patients with any of the series of skeletal dysplasias (chondrodystrophies) known to affect also the respiratory tract cartilages (see below), we know no data suggesting a specific and direct association of any of these tracheal structural patterns with respiratory tract disease.

### Epithelium

Tucker et al.[22] have described postnatal cephalad extension of ciliated respiratory epithelium over the subglottic region, laryngeal ventricles, lower laryngeal vestibule and portions of the cords. Whether this schedule is deranged in any specific diseases is apparently not known. Haddadin and Emery[23,24] demonstrated extensive 'squamous metaplasia' of the tracheal lining in 80% of 50 children dying with oesophageal atresia and tracheo-oesophageal fistula, the squamous epithelium being most marked over the posterior pars membrancea, but also extending to lateral and anterior walls and into the bronchi. They interpreted the change, not as reactive to infection, oxygen therapy etc., but as a component of the malformation complex, and proposed that the reduction in ciliary clearance of the respiratory tract caused liability to respiratory tract infection and/or to 'retention lung' in these infants. Similar effects of liability to infection and to sequelae thereof can perhaps be postulated for the 'immotile cilia syndrome', the significance of which as a general cause of tracheitis, bronchitis, bronchiolitis, broncho-pneumonia and otitis media in infants remains to be established. The association of ciliary anatomical and functional abnormality is best es-tablished for the Kartagener's syndrome of chronic respiratory tract infection, sinusitis and bronchiectasis, plus situs inversus in about half the patients[25]. At least seven different anatomical abnormalities of respiratory epithelial cilia have been described, so that the immotile cilia syndrome is clearly heterogeneous, and its full incidence equally clearly unknown. Described ciliary abnormalities include[25]:

(1) Lack of dynein arms of ciliary microtubules.
(2) Lack of parallel orientation of central ciliary microtubules.
(3) Displacement of central ciliary microtubules with lack of radial axoneme spokes.
(4) Compound cilia.
(5) Absent central microtubules.
(6) Absent subunits or doublets of peripheral microtubules.
(7) Translocation of peripheral microtubule doublets to central location.

Detailed analysis of the functional abnormality of the cilia (immotility versus abnormal direction of beat versus asynchrony etc.) in the light of the anatomical abnormality would undoubtedly improve understanding of the intracellular and intercellular control of ciliary activity. Appropriate study of whether some of these variations of ciliary structure are actually inherited by the mechanism of cytoplasmic inheritance would be of great interest.

The ultrastructural features of the endocrine (Kulchitsky) cells of the peripheral respiratory tract are surveyed by McDougall[26]; much more information is needed on functional, quantitative and cytological features of these cells in various diseases and malformations of the respiratory tract.

*Connective tissue septa*
These sheets of areolar tissue which extend into the pulmonary parenchyma from the pleura and from the connective tissue sheaths around bronchi and vessels appear at 18–20 weeks of fetal life[27]. Reid and Rubino found that the septa develop in the subpleural regions in relation to edges and angles of the lung, so that the larger flatter surfaces of the lung are relatively free of septa, except for the diaphragmatic surface, which has a higher number. They are thus found especially in the anterior edges of the upper and middle lobes, along the costophrenic edges and the costovertebral borders, and are rare over the lateral and costal aspects of the lungs. No explanation for this pattern has been proposed[27].

15

# SPECIFIC MALFORMATIONS OF THE RESPIRATORY TRACT

Available general reviews of respiratory tract malformations, variously covering clinical, anatomical (pathological) and radiological features of the many anomalies of these structures, and with widely varying content of pathogenetic information, include:

| reference | covers malformations of |
|---|---|
| 28 | nasal passages, pharynx, larynx, trachea, bronchi, lungs, diaphragm and chest wall, |
| 29 | larynx, trachea, bronchi, lungs, |
| 30, 31 | larynx, |
| 4, 32, 33 | trachea, bronchi, lungs, |
| 34, 35 | bronchi and lungs, |
| 36 | trachea, bronchi and lungs, associated with congenital cardiovascular anomalies. |

## Larynx

The larynx arises as the laryngotracheal groove, which develops as a median depression of the floor of the pharynx; its margins fuse from caudad forward to convert the groove into the laryngotracheal tube, which grows caudally to form the respiratory tract. The laryngeal aditus is temporarily obstructed by growth of the lateral epithelium, except for a small posterior canal, the ductus pharyngotrachealis. The epiglottis arises from the caudal hypobranchial eminence cephalad to the laryngotracheal groove[37].

### Laryngeal stenosis

Laryngeal webs, presumably remnants of the tissue which temporarily occludes the embryonic larynx, are, as would be expected, almost always anterior[31]. They may be associated with other anomalies, including tracheo-oesophageal fistula[38]. The larynx is abnormally narrow, and the epiglottis flaccid, so that it falls back to obstruct the laryngeal inlet and cause stridor, in the cri-du-chat (5p−) syndrome[39], and the epiglottis is flaccid but also hypoplastic in the Mohr syndrome of orofacial and digital anomalies and in the Majewski syndrome, which also includes skeletal dysplasia with short-limbed dwarfism and short-rib thoracic dystrophy, and urogenital tract anomalies (these two conditions may be versions of one basic disorder)[40]. Calcification present at birth of the hyoid, laryngeal and tracheal cartilages is reported in a small proportion (4 of 27 reported) of patients with chondrodysplasia punctata; airway obstruction may be laryngeal or tracheal[41]. Goldbloom and Dunbar, who may have described the same condition, reported slow decrease of the stridor over $2\frac{1}{2}$ years[42]. Laryngeal stenosis with apparent dominant inheritance pattern, without other overt anomalies, is also reported[43].

### Laryngeal atresia

This presumably reflects the same transient occlusion of the anterior laryngeal lumen by growth of the lateral laryngeal tissue which causes at least some forms of laryngeal stenosis. The locus of atresia may be both supra- and infraglottic (Type 1), infraglottic (Type 2), or glottic. Smith and Bain's data[44] on nine patients suggest that laryngeal atresia is a component of at least three different multiple anomaly syndromes.

(1) A syndrome of laryngeal atresia, hydrocephalus and vertebral anomalies (3 of 9 patients.).

(2) Laryngeal atresia as a component of the expanded anomaly syndrome associated with oesophageal atresia and tracheo-oesophageal fistula (EA–TEF) (the VATER, VACTEL, ARTICLE, ARTICLE-V etc. constellations) (4 of 9 patients.).

(3) Laryngeal atresia plus tracheal agenesis and broncho-oesophageal connection(s) (see below) (1 of 9 patients.).

### Laryngotracheo-oesophageal (LTE) cleft
This condition results from failure of the normal rostral extension of the tracheo-oesophageal septum, which separates the oesophagus from the developing respiratory tract; the cleft may involve the larynx only (10 of 22 patients reviewed by Gaskill and Bailey[45]), the larynx and upper trachea (6 of 22 patients) or the larynx and entire trachea (6 of 22 patients), the latter pattern also being called persistent oesophagotrachea[46]. Familial occurrence of laryngotracheo-oesophageal cleft is relatively common[47,48]; some of these reports refer to the 'G syndrome', which also includes facial abnormalities, hypospadias and oesophageal dysfunction[49]. Particularly associated with LTE cleft is oesophageal atresia, usually with tracheo-oesophageal fistula, which may be multiple[50,51], and often with other features of the EA–TEF anomaly complex (see below). Forrester and Cohen's report of three sibs with oesophageal atresia, anorectal anomaly and probable laryngeal fissure[52] raises the question of two aetiologically different mechanisms of association of LTE cleft with oesophageal atresia and/or tracheo-oesophageal fistula. Other respiratory tract anomalies associated with LTE cleft include subglottic stenosis[48] and pulmonary hypoplasia[53]. For unknown mechanisms, hydramnios is frequent in association with LTE cleft, even in the absence of oesophageal atresia[45,47,48,50]. (Note that hydramnios also occurs with laryngeal stenosis in the Mohr syndrome[40].)

## Trachea

As the laryngotracheal groove is separated from the oesophagus by the cephalad growth of the tracheo-oesophageal septum, it grows caudally as the primitive trachea, with the branched lower end forming the right and left lung buds[37].

### Congenitally short trachea
This anomaly occurs in the brevicollis syndrome[29,54].

### Tracheal stenosis
Tracheal stenosis may be diffuse or generalized (about 30% of patients with tracheal stenosis), funnel-like (about 20%) or segmental (about 50%), with

the locus of the segmental stenosis about evenly distributed in the subcricoid area, the central trachea and the supracarinal region[55].

Some instances of segmental tracheal stenosis have the condition called segmental oesophageal trachea, in which a segment of the trachea has the anatomical features of the oesophagus[56]. This is a cause of 'tracheomalacia' due to segmental absence of airway cartilage, and the example published by Landing and Wells[17] may well have had segmental oesophageal trachea.

### Tracheal stenosis with sling left pulmonary artery

In this complex, also called origin of the left pulmonary artery from the right, the left pulmonary artery arises farther to the right than usual and passes between the lower trachea and the oesophagus to the left lung, forming the arterial 'sling' around the supracarinal trachea[57]. The frequent (over 50%) association of tracheal stenosis due to complete (napkin-ring) tracheal cartilage rings (absence of the pars membranacea of the trachea) with sling left pulmonary artery is well reported (e.g., in references 11, 57–59). Sade et al.[59] explain the origin of the sling artery as follows: The developing lung bud has a blood supply from the 'pulmonary postbronchial plexus', a portion of the splanchnic vascular plexus. A ventral bud from the aortic sac on both right and left sides joins the postbronchial plexus to form the primitive pulmonary artery. If this ventral aortic bud does not develop, the lung may capture blood supply from other arteries via the postbronchial plexus. If this captured blood supply is from the area supplied by the right sixth aortic arch, as the developing lung moves caudad, the left pulmonary arterial branch of the right pulmonary artery will course behind the tracheobronchial tree. Namely, they propose that the primary defect in the sling left pulmonary artery anomaly is deficiency of the ventral bud off the aortic sac which should form the left pulmonary artery. Why this developmental aberration should so frequently be associated with abnormal growth and development of the tracheobronchial cartilages needs explanation. It seems clear that the ring tracheal cartilages do not reflect pressure on the trachea by the sling artery, since such cartilage lesions do not occur in association with the much more common (as well as more severe in terms of tracheal compression) syndromes of double aortic arch and other forms of tracheal constriction by aortic arch and great systemic artery anomalies (see below). Cohen and Landing[60] described abnormal cartilage patterns in the main bronchi in 2 of 3 patients with sling left pulmonary artery studied by them, further indicating that compression by the sling artery is not the cause of the cartilage abnormalities[61]. The 'pseudoisomerism' of the main bronchi reported by these workers is presumably at least partially explained by elongation of the right main bronchus, over which the left pulmonary artery must pass (compare the relatively long right main bronchus often seen with right aortic arch)[61].

### Ductus arteriosus sling

Binet et al.[62] reported an instance of course of the ductus arteriosus behind the carina from the aorta to the right pulmonary artery. They explained this anomaly also as development of a connection from the left sixth arch to the right through the postbronchial plexus between the trachea and oesophagus, which normally regresses at the 8 mm embryonic stage.

### Tracheomalacia

This term originally implied tracheal narrowing or collapsing due to abnormally soft and pliable tracheal cartilages, but this is apparently rarely the fact. Wittenborg et al.[63] demonstrated that most patients with inspiratory tracheal collapse at the cervicothoracic junction actually were demonstrating the normal response of the trachea to abnormally great negative pressure in the lower airway as the result of high airway obstruction, and that patients with more diffuse expiratory collapse of the intrathoracic trachea were demonstrating the normal effect of increased expiratory pressure due to small airway obstruction, as in asthma or bronchiolitis. Tracheal flattening or narrowing, abnormal tracheal cartilage patterns, abnormal bronchial branch patterns, or combinations of these have been reported in a number of skeletal dysplasia or 'chondrodystrophy' syndromes, including chondrodysplasia punctata[41], Ellis – Van Creveld syndrome[17,64], camptomelic dwarfism[17], Jeune's thoracic asphyxiating dystrophy[29], diastrophic dwarfism[17], thanotophoric dwarfism with clover-leaf skull[17], and Saldino–Noonan syndrome[29]. The syndrome, reported by Smith et al.[65], of abnormally flat trachea (but with narrow rather than wide pars membranacea), rib-gap defects, micrognathia and redundant skin is another 'tracheoskeletal' syndrome.

### Tracheal compression by aortic and arterial malformations

An excellent review of this group of entities, with explanation of the embryological mechanism of each, is that by Edwards[66]; other useful reviews include those by Gross[67], Mustard et al.[68], Lincoln et al.[69] and Fearon and Shortreed[70]. Vascular anomalies in this category include:

*Relative numbers in these series*

17     (1) Anomalous (retro-oesophageal) right subclavian artery.

70     (2) Double aortic arch, with or without coarctation of one or both arches, with or without segmental atresia of the left arch, and with or without abnormally short ductus arteriosus.

25     (3) Right aortic arch with retro-oesophageal segment and left ductus arteriosus.

     (4) Left aortic arch with retro-oesophageal segment and right ductus arteriosus.

     (5) Right aortic arch with aberrant left subclavian artery.

     (6) Interruption of the aortic arch with right descending aorta and right ductus arteriosus.

89        (7) Anomalous origin of the brachiocephalic (innominate) artery.

(8) Anomalous origin of the left common carotid artery.

Compression of the lower left trachea by the ductus arteriosus in a patient with congenital absence of the right lung[71], by the left pulmonary artery after right pneumonectomy[72], and by the left aortic arch after right pneumonectomy[73] have also been reported.

*Tracheal compression by arterial anomalies in patients with oesophageal atresia and tracheo-oesophageal fistula (common origin of the right and left carotid arteries)*

As is shown above, tracheal compression by anomalous brachiocephalic or left common carotid arteries is the single most common cause of tracheal compression by aortic arch branches; this conclusion is confirmed by the recent data of Idriss *et al.*[74]. Although variously described as 'origin of the brachiocephalic artery farther to the left than normal'; or as 'origin of the left common carotid artery farther to the right than normal', a more apt general description of the anomaly would appear to be 'common origin of the right and left carotid arteries' – of the left common carotid artery with the brachiocephalic when the right subclavian artery has a normal origin, and of the left common carotid artery with the right when there is an anomalous right subclavian artery[75].

Although sometimes attributed to cartilage deficiency, possibly somehow due to pressure on the trachea by the dilated oesophageal segment[76], tracheal narrowing in patients with oesophageal atresia and tracheo-oesophageal fistula (EA–TEF) is also ascribed to pressure on the trachea by abnormal arteries[7,7,78]. To investigate this matter, we have reviewed the gross specimens from 25 patients with oesophageal atresia and tracheo-oesophageal fistula, and of one patient with 'H-type' tracheo-oesophageal fistula, available in the files of the Department of Pathology, Childrens Hospital of Los Angeles. The relevant data are given in Table 2.5, which presents the location of the aortic arch and of the descending aorta, the type of great artery anomaly present, whether the right main and right upper lobe bronchi were abnormal, whether the trachea showed a groove from pressure by an abnormal artery, whether other bronchial anomalies or congenital heart disease were present, the location of the ductus arteriosus, and the nature of other malformations found.

The data of Table 2.5 on abnormal great arterial patterns in patients with EA–TEF are summarized in Table 2.6, in comparison with comparable data from series of patients with complete transposition of the great arteries, with hypoplastic left heart complex, and with normal hearts. The high incidence of common origin of the left common carotid artery with the right brachiocephalic or common carotid artery in patients with EA–TEF confirms the suggestions, given above, that tracheal compression by abnormal arteries is frequent with EA–TEF (note that such tracheal compression is perhaps exaggerated by the tracheal cartilage hypoplasia seen with EA–TEF, as described above[19]). Pathologically, common origin of the left carotid and

**Table 2.5** Pathological findings with oesophageal atresia and tracheo-oesophageal fistula or isolated ('H-type') tracheo-oesophageal fistula

| Age | Sex | Arch | Desc. aorta | Great artery abnormality | RMB/RULB abn. | Tracheal groove | Other bronchial abnormal. | Cardiac malf. | DA | Other malformations |
|---|---|---|---|---|---|---|---|---|---|---|
| *EA–TEF* | | | | | | | | | | |
| 1. 2 d | M | L | L | 0 | above carinal level | +lt | 0 | + | L | Imperf. anus; rectourethral fistula; VSD, PS |
| 2. 3 d | F | R | L | order RSC, RCC, LBC | carinal level | +rt +lt | — | + | 0 | ASD, VSD, PS; horseshoe kidney; vesicovaginal fistula; bifid uterus; 'situs inversus' heart |
| 3. 3 d | M | ?L | ?L | 0 | carinal level | — | 0 | — | 0 | imperf. anus; rectourethral fistula; crossed renal ectopia with hypoplasia; phocomelia, arms; choanal stenosis, lt |
| 4. 3 d | M | ?R | 0 | common RBC +LCC | long down-curved RMB | +lt | trachea curved to left | + | ?L | VSD; hydrocephalus; malrotation; imperf. anus; horseshoe kidney; vertebral anom. |
| 5. 3 d | F | L | L | common RBC & LCC | high origin of RULB | +lt | (? cart. def. lt lung trachea) | + | L | Hypoplastic lt heart complex, VSD, AS, PDA, probable 18 trisomy |
| 6. 4 d | F | ?L | ?L | 0 | carinal level | +lt | LMB short (? RAA) | — | ?L | — |
| 7. 4 d | F | L | L | RBC & LCC abn. close | 0 | +lt | 0 | + | ?L | Triad syn.; hydroureter & renal dysplasia; malrotation; phocomelia, lt arm; VSD; vert. anom. |
| 8. 4 d | M | L | L | common RBC & LCC | carinal level | 0 | — | — | L | — |
| 9. 5 d | M | ?L | ?L | 0 | carinal level | +lt | horizontal LMB | — | ?L | — |

(continued)

21

**Table 2.5** (continued)

| Age | Sex | Arch | Desc. aorta | Great artery abnormality | RMB/RULB abn. | Tracheal groove | Other bronchial abnormal. | Cardiac malf. | DA | Other malformations |
|---|---|---|---|---|---|---|---|---|---|---|
| *EA–TEF* | | | | | | | | | | |
| 10. 5 d | M | L | L | common RBC +LCC | 0 | +lt | situs inversus hypopl. lt lung | – | L | mult. rib & vertebral anomalies; absent left radius & thumb; hypoplastic, left lung; single kidney with pelvic ectopia |
| 11. 5 d | F | L | L | RBC & LCC abn. close | carinal level | 0 | – | + | L | VSD; imperforate anus with rectovaginal fistula; horseshoe kidney; vertebral abn. |
| 12. 6 d | M | L | L | 0 | +lt | 0 | 0 | – | L | megaureter, bilateral |
| 13. 12 d | F | L | L | – | carinal level RULB | – | – | – | L | Malrotation |
| 14. 2 w | M | ?R | 0 | 0 | long RMB with down-curved RULB | +lt | – | – | ?L | – |
| 15. 16 d | F | L | L | RBC & LCC abn. close | carinal level | +lt | horizontal LMB | + | L | OAVC; DORV; PS; LSVC; malrotation; urethral stenosis with hydroureter & hydronephrosis; Meckel's diverticulum; syndactyly; trisomy 18 |
| 16. 3 wk | M | L | L | 0 | 0 | +lt | 0 | + | absent | arhinencephaly; VSD, LSVC, truncus type 1; cystic dysplasia, lt kidney; bilat. absence, radii & thumbs |
| 17. 3 wk | F | L | L | – | 0 | +lt | 0 | + | L | Kallmann & DiGeorge syndromes; hypoplastic left heart complex + VSD; cleft palate; tracheal groove? |
| 18. 1 mth | M | R | L | retro-oesoph. LSC | nl. | – | LMB short | – | L | imperf. anus; ureteral atresia & dysplasia, kidney, rt |

22

| No. | Sex | Arch | DA | Great artery abnormality | RMB/RULB abnormality | Tracheal groove | Other bronchial abnormalities | PDA | DA | Other malformations |
|---|---|---|---|---|---|---|---|---|---|---|
| 19. 1 mth | M | L | L | — | 0 | 0 | — | — | L | — |
| 20. 9 wk | F | L | ?L | common RCC & LCC anomalous RSC | 0 | +lt | 0 | — | 0 | horseshoe kidney; pyloric stenosis; foramen of Morgagni hernias, diaphragm; trisomy 18 |
| 21. 9 wk | M | ?L | ?L | — | ? nl. | 0 | — | — | ?L | Down syndrome |
| 22. 4 mth | M | L | L | RBC & LCC abn. close | ? high origin RULB | 0 | — | — | L | omphalocoele, malrotation, horseshoe kidney |
| 23. 6 mth | M | L | L | common RBC & LCC | marked distal displacement | diffuse stenosis | LMB short | — | L | laryngeal stenosis; sling LPA with tracheal stenosis with complete cartilage rings; cystic dysplasia, rt kidney; bronchial branch pattern abnormal bilaterally, with 3 RUL bronchi and 3 left lobar bronchi with bronchus intermedius |
| 24. 9 mth | M | L | L | common RBC + LCC | carinal level | +lt | horizontal LMB | + | ?L | imperf. anus; VSD; absence rt kidney; malrotation; malf. thumbs |
| 25. 3.5 y | F | L | L | common.RCC & LCC; anomalous RSC | nl. | +rt | — | + | L | VSD, MS; autopsy restricted to thoracic organs |

TEF

| No. | Sex | Arch | DA | Great artery abnormality | RMB/RULB abnormality | Tracheal groove | Other bronchial abnormalities | PDA | DA | Other malformations |
|---|---|---|---|---|---|---|---|---|---|---|
| 1. 5 mth | F | L | L | common RBC & LCC | carinal level RULB | +lt | — | — | L | H-type TEF; — |

B

*Arch: Aortic arch was left (L) or right (R); 0 = locus of arch could not be determined from specimen. Great artery abnormality: RSC = right subclavian artery; RCC = right common carotid artery; LCC = left common carotid artery; RBC = (right) brachiocephalic artery; LSC = left subclavian artery; 0 = arterial pattern could not be established from specimen. RMB/RULB abnormality: RMB = right main bronchus; RULB = right upper lobe bronchus; 0 = pattern could not be determined from specimen. Tracheal groove: + = groove present; − = groove not present; lt, rt = groove on left or right. Other bronchial abnormalities: LMB = left main bronchus; 0 = pattern could not be established from specimen. Other malformations: VSD = ventricular septal defect; PS = pulmonic stenosis; PDA = patent ductus arteriosus: L = left ductus arteriosus; 0 = locus of DA could not be established; DA = ductus arteriosus; DORV = double outlet right ventricle; OAVC = ostium atrioventricular commune; LSVC = left superior vena cava; MS = mitral stenosis.*

**Table 2.6**

| No. Adequate specimens | Oesophageal atresia/tracheo-oesophageal fistula 17 | | Transposition of great arteries 25 | | Hypoplastic left heart complex 20 | | Control series 20 | |
|---|---|---|---|---|---|---|---|---|
| | No. | % | No. | % | No. | % | No. | % |
| Common origin | 8 | 47 | 1 | 4 | 1 | 5 | 2 | 10 |
| Close origin | 4 | 24 | 9 | 36 | 5 | 25 | 7 | 35 |
| 'Normal' relation | 4 | 24 | 14 | 55 | 9 | 45 | 11 | 55 |
| Other vascular anomaly | 1 | 6 | 1 | 4 | 5 | 25 | 0 | 0 |
| | | | (anomalous right subclavian art.) | | (abn. large rt. brachiocephalic art.) | | (abn. large rt. brachiocephalic anomalous rt. − 4 subclavian art. − 1) | |

Incidence of common origin of left common carotid artery with (right) brachiocephalic artery or right common carotid artery, of close origin of LCC with RBC or RCC, of most common or 'normal' relation of origins, and of other arch vessel anomalies in patients with oesophageal atresia and tracheo-oesophageal fistula, in comparison to patients with transposition of great arteries, with hypoplastic left heart complex, and with normal hearts. The data demonstrate: (1) The high incidence of common origin of the left common carotid artery with the right in patients with EA–TEF, in comparison with patients with common congenital heart lesions typically fatal in early infancy (TGV, HLH). (2) The most common or 'normal' relation of the ostia of the LCC with the RBC or RCC is found in about 50 % of normals, TGVs and HLHs. (3) Close (but not common) origin of LCC with RBC or RCC is a frequent normal variant, occurring in 25–35 % of patients studied

(right) brachiocephalic arteries is most easily confirmed by opening the aorta, and observing only two main arterial ostia (unless there is also anomalous (retro-oesophageal) right subclavian artery) (Figures 2.2 and 2.3). The term in Table 2.6, 'close origin' of the left carotid and brachiocephalic artery, refers to separation of the appropriate ostia by a narrow V-shaped septum. This anatomical variation is clearly frequent, and not selectively associated with EA–TEF, transposition of the great arteries or hypoplastic left heart complex. Clinical experience suggests that differentiation of these two situations by angio-cardiography can be difficult.

*Embryology of common origin of right and left common carotid arteries*
The standard Meckel (or Rathke) diagrams of the development of the human aorta and great arteries from the embryonic branchial arch vessels, as presented in many text books of embryology, or other diagrammatic presentations of normal and abnormal developmental patterns of these vessels, would appear to indicate that common origin of the right and left common carotid arteries from a single orifice from the aortic arch is embryologically impossible – the right and left common carotid arteries each arising from the corresponding ventral aortas into which the common ventral aortic trunk divides at the level of the fourth or fifth aortic arches[66,79–81]. That this is not the case is clearly shown by the material presented above. The scheme of aortic arch development shown in Williams·et al.[37] approximates

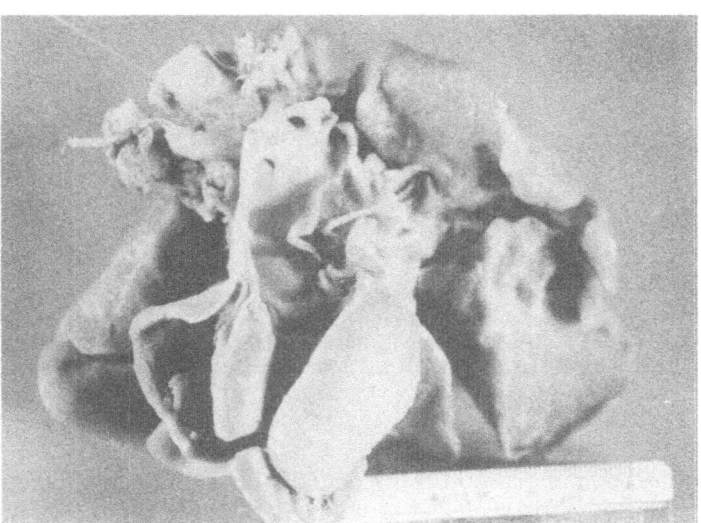

**Figure 2.2** A, Specimen of patient with EA–TEF, showing common origin of the right and left common carotid arteries from a short trunk vessel, with the trachea held in the V-shaped notch between the two common carotid arteries. B, View of same specimen from the inner aspect of the aorta, showing the single ostium for the two common carotid arteries

the condition called (above), 'close origin' of the right and left carotid arteries, and that given in Thomas[82] is the only one we have seen which clearly presents an embryonic stage with common origin of the right and left common carotid

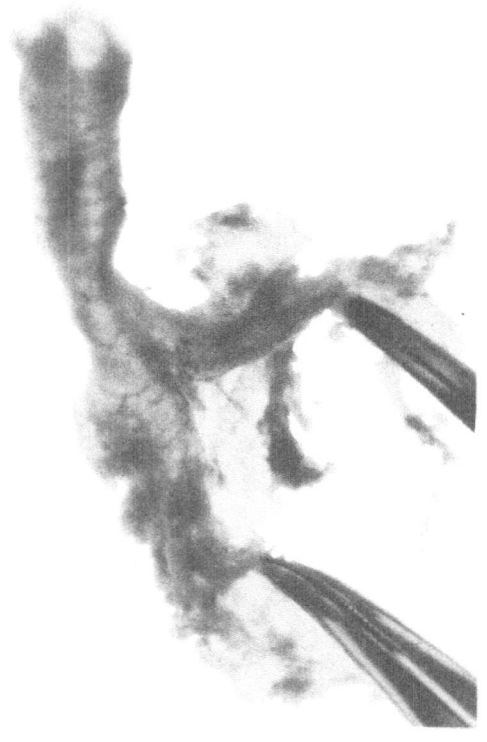

**Figure 2.3** Lateral view of dissected lower airway of infant with EA—TEF and common origin of the common carotid arteries, showing the tracheal notch caused by the pressure of the left common carotid artery on the left lower anterior aspect of the trachea

arteries, the common carotid trunk actually being the continuation of the primitive ventral aorta cephalad to the level of the origin of the left fourth aortic arch (the definitive human aortic arch) and the right fourth arch (the right subclavian artery), with the ventral aorta not dividing until the level of the second aortic arches. Whether this scheme reflects a developmental stage found in all normal human embryos, or a pattern highly associated with EA—TEF but also found in a small proportion (4–10% in the data of Table 2.6) of patients with other disorders, and hence within the 'range of normal variation', we do not know (Fig. 24).

## Tracheal compression by abnormal arteries in patients with the hypoplastic left heart complex

In the review of aortic arch and great artery patterns in patients with EA–TEF, the data of which are presented above, hypoplastic left heart complex (HLH) was employed as a 'control' disorder for which adequate material was available in our Cardiac Registry. It is of interest that all patients with this disorder for whose specimens the location of the aortic arch could be established (19 of 20) had a left aortic arch (this was also true for transposition of the great arteries), in contrast to the 10 % (and possibly 15 %) incidence of right aortic arch in the patients with EA–TEF listed in Table 2.5. The increased association of right aortic arch with EA–TEF, and its implications as regards surgical repair of the condition, are discussed by Harrison et al.[83]. The findings in 20 patients with HLH are summarized in Table 2.7, which shows that a tracheal groove by the (right) brachiocephalic artery was seen in 6 of 18 specimens adequate for study of this point, and that for 4 of these 6 specimens the brachiocephalic artery appeared disproportionately large. Preferential flow of the small amount of blood coming up the ascending aorta in patients with HLH into the brachiocephalic artery, so that only the right arm region and the right half of the face and head appear pink, with the rest of the body cyanotic, is a feature of HLH, so that disproportionately large (right)

**Table 2.7**

|  | Great artery pattern | Aortic arch | Tracheal GV groove | Tracheal DA groove |
|---|---|---|---|---|
| 1. | Normal | L | – | – |
| 2. | N | L | – | +anterior |
| 3. | N | L | – | – |
| 4. | N | L | – | +ant. left (aorta) |
| 5. | N | 0 | 0 | 0 |
| 6. | N | L | – | ?ant. left (aorta) |
| 7. | N | L | – | +anterior |
| 8. | N | L | – | ?anterior |
| 9. | N | L | – | ?ant. left (aorta) |
| 10. | ⎫ | L | – | ?anterior |
| 11. | ⎪ | L | – | +anterior |
| 12. | ⎬ Close origin LCC & RBC | L | 0 | 0 |
| 13. | ⎪ | L | + rt. (RBC) | + left (aorta) |
| 14. | ⎭ | L | + rt. (RBC) | + anterior |
| 15. | Close origin LCC & RBC, RBC large | L | + rt. (RBC) | + anterior |
| 16. | Common origin, LCC & RBC | L | + rt. | + anterior |
| 17. | N, RBC large | L | + rt. (RBC) | + anterior |
| 18. | N, RBC large | L | + rt. (RBC) + lt. (LCC) | ?anterior |
| 19. | N, RBC large | L | + rt. (RBC) | + left (aorta) |
| 20. | Anomalous RSC | L | – | ?anterior |

Great vessel and tracheal groove patterns, patients with hypoplastic left heart complex, showing association of a right tracheal groove with a disproportionately large (right) brachiocephalic artery, and the high incidence of left anterior tracheal grooving by the aorta in the isthmic region. The large ductus arteriosus seen in the hypoplastic left heart complex, directed anteroposteriorly, anterior to the aortic isthmus and medial to the left main bronchus at the carina, presses the aorta against the lower trachea brachiocephalic artery (? really disproportionately small left common carotid and subclavian arteries) in a significant proportion of patients with HLH is not unexpected. Whether the relatively large patent ductus arteriosus, which in patients with HLH supplies the lower body with arterial blood, compresses the lower trachea or left main bronchus was investigated. The ductus arteriosus, which is directed antero-posteriorly anterior to the aortic isthmus, lies basically medial to the left main bronchus near the carina, and does not arch over the left main bronchus, but the findings suggest that the large ductus in HLH 'presses' the aortic arch against the left anterior aspect of the lower trachea more than does the normal ductus arteriosus (although whether this is ever of symptomatic significance cannot be stated from this material).

### Tracheal agenęsis (absence, aplasia)

This condition is another example of the failure of the developmental schemes presented in standard embryological sources to explain fully the genesis of relatively common malformations. Three main forms of tracheal agenesis occur[84].

(1) *Type 1*: Absence of the proximal trachea with presence of the distal trachea, carina and bronchi.

(2) *Type 2*: Absence of the trachea to the carina, with the main bronchi joining in the midline and communicating with the oesophagus by a common fistulous tract.

(3) *Type 3*: Absence of the trachea and carina, with the right and left main bronchi communicating independently with the oesophagus.

Collectively, tracheal agenesis shows modest male preponderance[85], and hydramnios may occur, as with laryngotracheo-oesophageal cleft[86]. Other respiratory tract anomalies which may occur in patients with tracheal agenesis include laryngeal atresia and unilateral or bilateral absence of the lungs[87].

According to standard descriptions of the development of the respiratory tract, as presented above, the caudal end of the laryngotracheal groove, which forms as a median longitudinal depression of the pharyngeal floor, is separated from the oesophagus by cephalad growth of the tracheo-oeso-phageal septum to form the tracheal bud, which grows caudally and branches at its tip to form the right and left main bronchi. This schema makes origin of a bronchus, or of the lower trachea, from the oesophagus difficult to explain, and what amounts to the proposition of the occurrence of an accessory or supernumerary laryngotracheal groove or lung bud, caudal to the normal groove, has been advanced to explain certain anomalies (see discussion of sequestration below)[88,89]. Perhaps more easily understood as a mechanism of origin of bronchi from the oesophagus would be the appearance of the two main bronchial buds from the pharyngeal floor before the tracheo-oeso-phageal septum separates the ventral pharyngeal wall (the laryngotracheal groove) from the oesophagus to form the trachea. Misplacement or diagonal alignment of the 'cutting edge' of the tracheo-oesophageal septum would then leave one bronchus attached to the trachea and one to the oesophagus. If the septum did not advance at all, both bronchial buds would remain attached to the oesophagus ( = tracheal agenesis Type 3), and if it advanced only part way to the larynx, the lower trachea (bearing the bronchi and lungs) would remain attached to the oesophagus ( = tracheal agenesis Type 2). In both these latter circumstances, as well as in tracheal agenesis Type 1, where the cephalic portion of the trachea is lacking, but the distal portion and bronchopulmonary structures are present (but not connected to the oesophagus), a process causing regression of the cephalic portion of the trachea appears necessary to explain the difference between tracheal agenesis and laryngo-tracheo-oesophageal cleft (see above); namely, tracheal agenesis Type 1 would be the result of the action of this 'tracheal regression process', tracheal agenesis Type 2 of the joint effects of the 'tracheal regression process' and of reduced growth of the tracheo-oesophageal septum, and tracheal agenesis Type 3 of the 'tracheal regression process' plus lack of cephalic growth of the tracheo-oesophageal septum (with the bronchial buds arising from the pharyngeal floor before the tracheal bud is separated from the oesophagus). At this moment, it appears unwise to ignore what would appear to be adequate evidence that in at least some humans the caudal tracheal bud is separated from the pharynx—oesophagus before it branches to form the primordial main bronchi, and to propose that the 'normal' process in humans is origin of the bronchial buds from the ventral pharynx before the laryngotracheal groove is separated from the oesophagus, but the proposition that this latter sequence occurs in some humans appears tenable. Of great interest in this regard is the report of Spooner and Wessells[90] that the normal sequence in development of the respiratory tract in mice is origin of the lung buds from the anterior pharynx before the tracheal anlage is separated.

The basic studies of Wessells and various co-workers[90-92] are of great interest as regards the pathogenesis of a variety of tracheo-broncho-pulmonary anomalies. These workers conclude, from transplantation studies on mouse embryo explants, that there are several levels of mesodermal control of lung development.

(1) Non-specific mediastinal mesoderm interacts with embryonic gut endoderm to cause lung (tracheal) bud formation, but inhibits bronchial branching.

(2) Specific bronchial mesoderm interacts with lung (tracheal) bud endoderm to produce bronchial branching and peripheral airway morphogenesis.

Thus, for instance, pulmonary agenesis (see below) could be the result of deficiency of specific bronchial mesoderm or of an excess of tracheal mesoderm, and supernumerary lung buds (or primary origin of two bronchial buds before the tracheal bud, as discussed above) of the presence of bronchial mesoderm adjacent to the ventral pharynx etc.


### Intratracheal thyroid

The thyroid primordium arises as an endodermal bud from the midline anterior pharyngeal floor between the first and second pharyngeal pouches. It extends caudally, superficial (anterior) to the laryngotracheal tube, as the thyroglossal duct, which normally involutes completely[37]. Intratracheal (rarely intrabronchial) thyroid is a relatively frequent incidental microscopic finding in infants; it may or may not communicate with adjacent thyroid gland through the tracheal wall, or may occur at levels superior or inferior to the normal thyroid. In view of the normal location of the thyroid gland and thyroglossal duct anterior to the developing respiratory tract, the apparent predilection of intratracheal thyroid tissue (or at least of symptomatic intratracheal goitre) to occur in the left posterolateral region of the trachea is difficult to explain[93] – if a thyroid remnant could somehow become attached to the inner end of a thyroid cartilage it could possibly be carried, by the normal posterolateral growth of the cartilages, to a posterolateral locus in the trachea, but why especially on the left side?

### Tracheo-oesophageal fistula without oesophageal atresia
The so-called H-type tracheo-oesophageal fistula (which should perhaps be called 'N-type', because the fistulous tract typically runs upward from oesophagus to trachea[94]), may be associated with segmental tracheal stenosis[95], which is at least sometimes due to the lesion described above as 'segmental oesophageal trachea'[56]. The relative rarity of other congenital anomalies occuring in association with this lesion, which may not be diagnosed until later childhood or adult life, suggests that it is not fundamentally related pathogenetically to the type of tracheo-oesophageal fistula associated with oesóphageal atresia, discussed below. That this type of TEF is multiple in an appreciable number of patients[96] could support a proposition that the defect involves abnormality of the tracheo-oesophageal septum as it pushes cephalad between the oesophagus and the developing trachea. Isenberg and Tubergen[97] reported a patient with two H-type fistulas who also had unilateral complete aplasia of the inner ear. They suggested that this association could be explained by a primary abnormality of the ventral archenteron which is the primary inducer of the otic vesicle and has a major inductive influence on the formation of the trachea.

### Tracheo-oesophageal fistula with oesophageal atresia (EA–TEF)
The expanded (and still expanding) list of other congenital anomalies seen in association with EA–TEF have received much attention in the recent literature[36,98–102]. Acronymic terms for these constellations have included VATER (vertebral, anal, tracheo-oesophageal, renal and radial), VACTEL (vertebral, anal, cardiac, tracheo-oesophageal, limb), ARTICLE and ARTICLE-V (anal, renal, tracheal, intestinal, cardiac, limb, oesophagus etc., vertebral, vascular). Although the coincidence of oesophageal atresia and tracheo-oesophageal fistula with congenital heart disease, aortic arch and great artery abnormalities (presented above), other respiratory tract malformations, and cervicodorsal vertebral and arm anomalies could indicate a 'field defect', the high incidence of ano-rectal, genitourinary and gastrointestinal tract lesions in these patients indicates that such an explanation of this anomaly complex is not adequate. Barry et al.[98] state that the complex should be regarded as 'a spectrum that ranges from the occurrence of any one of the defects singly to the full set' of anomalies, a believable situation but one for which it is very difficult to propose a biochemical pathogenetic mechanism. Whether the presence of certain anomalies (e.g., spine or rib[99]) in patients

who fall into this complex significantly indicates the presence of others has been debated, but it is still not certain that the view that the predictive power of the VATER etc., concept is greater than that of the general risk of occurrence of each of the anomalies independently in patients falling in the group. Useful data on other respiratory tract anomalies in patients with EA–TEF are contained in the series of German et al.[100] and Toyama[101] (and see also above, under *laryngeal atresia, laryngotracheo-oesophageal cleft* and *tracheal compression*, and below, under *pulmonary agenesis*). That a biochemical interaction between urinary tract and respiratory tract, of the sort proposed by Hislop et al.[102] to explain the pulmonary hypoplasia of the Potter syndrome can apply to the coincidence of kidney and urinary tract anomalies with EA–TEF seems unlikely, since EA–TEF is, by definition, the necessary and sufficient criterion for membership in the complex, not nephrourinary tract lesions (48% in Atwell and Beard's data[103], 12% in reference 100, 19% in reference 99 and at least 56% in the data of Table 2.5).

EA–TEF is apparently a less rigidly stereotyped multiple anomaly complex than the Arnold–Chiari neural tube defect complex or the Ivemark asplenia syndrome, two other common and significantly lethal entities which are, like EA–TEF, desperately in need of pathogenetic information and imaginative pathogenetic hypotheses. As is implied above, it is difficult to formulate hypotheses on the causes and mechanisms of a condition that at present appears to be the biological equivalent of being hit by buckshot. Warren et al.[104] studied the incidence of EA–TEF in the siblings and children of 79 affected persons surviving to child-bearing age; of 130 sibs, 1 also had EA–TEF, and of 28 children, 1 was affected. This recurrence is too low for EA–TEF to be generally the result of a new dominant mutation which was lethal before the modern surgical period.

## Abnormal bronchial branching patterns

### Development of the main and lobar bronchi
The caudal tip of the tracheal end of the respiratory diverticulum, which forms as the caudal end of the laryngotracheal groove separates from the pharyngo-oesophageal tube, is usually stated to bifurcate to form the right and left main bronchi by the end of fetal week 4[82]. The stem bronchus of each lung branches to give rise to dorsal and ventral branches. The right upper lobe bronchus probably actually arises as a separate dorsal bud, originally above the carinal level, rather than truly off the right stem bronchus[81]; the relevance of this point to the bronchial branch pattern called 'tracheal trifurcation' is discussed below. The pulmonary artery on the right thus originally runs ventral to the upper lobe (apical) bronchus and dorsal to the middle lobe bronchus, whereas the left pulmonary artery runs dorsal to the main bronchus[105]. The actual sequence of branching to form the lobar and segmental bronchi can be inferred from the diagram given by Bucher and Reid[4].

### Pulmonary isomerism patterns
Described patterns of pulmonary isomerism (the presence of the same lobar branch pattern in both lungs) include the following[106].

(1) The Ivemark asplenia syndrome with pulmonary isomerism of bilateral right lung type. This condition – which includes congenital heart disease (typically absence of the ventricular septum, atrial septal defect, transposition of the great vessels, pulmonic stenosis or atresia, total anomalous pulmonary venous return and abnormal great systemic vein patterns), abnormal intestinal rotation and symmetry of the liver, in addition to the pulmonary isomerism with right lung pattern bilaterally and absence of the spleen – has been proposed to be due to 'bilateral right-sidedness'. The condition shows male preponderance, and is not apparently genetically determined.

(2) Polysplenia, with pulmonary isomerism of bilateral left lung type, congenital heart disease (typically, septal defect), abnormal intestinal rotation, symmetric liver, abnormal great vein patterns and multiple small spleens. This syndrome has been proposed to result from 'bilateral left-sidedness'. Sex ratio in polysplenia is equal, and it is also not apparently a genetic disease.

(3) M-anisosplenia, the least well defined of these four concepts, consists of pulmonary isomerism of right lung type plus congenital heart disease. Visceral situs is normal, and the spleen consists of non-uniform small spleens ('one or more larger and one or more smaller'[106]).

(4) F-anisosplenia consists of pulmonary isomerism of bilateral left lung type, congenital heart disease (most typically, double outlet right ventricle), intestinal malrotation, and anisosplenia as described under pattern (3). All patients with this syndrome we have seen to date have been female.

(5, 6) There is suggestive evidence of the occurrence of pulmonary isomerism of right lung type in the Ellis-van Creveld syndrome[17], and in bronchobiliary fistula[33,107]. The incidence of these associations is not known.

(7) Whether bronchial isomerism is present in the interesting familial syndrome of hypoplastic left heart complex, bilateral nonlobation of the lungs, polydactyly, and other anomalies, plus hydramnios, reported by Kohler[108], is not known. Review of the records and specimens of 136 patients with hypoplastic left heart complex in the Cardiac Registry, Childrens Hospital of Los Angeles, disclosed no other patients with this pattern of anomalies.

*Bridging bronchus (? two different syndromes)*
As described by Gonzalez-Crussi *et al.*[109], in bridging bronchus, the right main bronchus supplies the right upper and middle lobes, while the bronchus from right lower lobe rises from the left main bronchus and 'bridges' the mediastinum. A similar anomalous bronchial pattern occurs in association with a form of sling left pulmonary artery[59,110]. In one such patient we have seen, the right bronchus supplied only the right upper lobe, and the left main bronchus gave rise to bronchi to the right middle and lower lobes. The literature suggests that this is the usual pattern when this form of sling left pulmonary artery is present (whence the occasional application of the term

32

'bronchus suis' to this complex), although more data are needed. If this is so, one must recognize three separate entities.

(1) Classical sling left pulmonary artery, with the artery passing behind the supracarinal trachea.

(2) Bridging bronchus plus sling left pulmonary artery, with the artery passing behind the left main bronchus, which supplies bronchi to the right middle and lower lobes.

(3) Classical bridging bronchus, with the left main bronchus furnishing the bronchus to the right lower lobe.

### Congenital bronchobiliary fistula

Although this anomaly was originally described as congenital tracheobiliary fistula, the fistula typically arises from a main bronchus, more commonly the right, near the carina[111,112]. The occasional association of other anomalies, including oesophageal atresia and limb malformations[113] may suggest a basic relation to the EA–TEF complexes discussed above, but the point has been made that no reported case has shown vertebral anomalies[111]. High origin of the right upper lobe bronchus, producing a pattern which can be called carinal trifurcation, reported in a patient with bronchobiliary fistula[114], may support a pathogenetic relation to EA–TEF (see next section).

### Bronchial branch pattern abnormality in patients with EA–TEF

Table 2.5 shows that 12 of 19 patients with EA–TEF, for whom the bronchial branch pattern could be established from specimens in the Cardiac Registry, Childrens Hospital of Los Angeles, had the right upper lobe bronchus arising at the level of the carina ( = cardinal trifurcation) or higher (tracheal bronchus) (Figure 2.5). Data that this pattern is statistically abnormal are given in Table 2.1 and Figure 2.1. That the pattern represents persistence of an early fetal bronchial pattern[81] can be proposed, but it is not associated with other multiple anomaly complexes, such as the Ivemark asplenia or the polysplenia complexes[106], so that relatively non-specific explanation of this type seems suspect (see also next section).

### Short right main bronchus in transposition of the great arteries (TGV)

Table 2.8 demonstrates that an abnormally short right main bronchus is frequent in patients with TGV (at least 10 of 23), further indicating that the association of the carinal trifurcation pattern discussed above with EA–TEF has specific significance (Fig. 2.6).

### Origin of a main bronchus from the oesophagus

This condition is sometimes included in the broad category of sequestration[115], but would appear to represent, at least in part, a different condition (although see discussion of sequestration, below). Bowen and Parry[116] state that the literature to date contains 11 reports of 'sequestration of the entire right lung' and none of the left lung (whereas typical extralobar and intralobar sequestrations occur predominantly on the left). Other

33

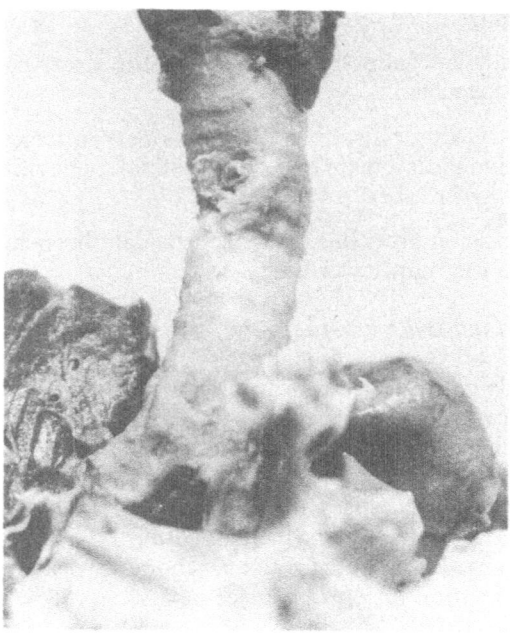

**Figure 2.4** Anterior view of trachea of patient with EA–TEF and common origin of right and left common carotid arteries, showing diagonal keel-like ridge of anterior trachea between the higher right and lower left notches caused by right and left common carotid arteries respectively. Note that the tracheostomy is above the site of tracheal compression by the anomalous arteries

**Table 2.8 High incidence of abnormally short right main bronchus in patients with complete transposition of great arteries (without Ivemark asplenia syndrome or other conotruncal anomalies)**

| | |
|---|---|
| Total specimens examined | 27 |
| Right main bronchus normal | 10 |
| RMB short | 10 |
| RMB? short | 3 |
| Status of RMB not determinable | 4 |
| Aortic arch left | 24 |
| Aortic arch right | 0 |
| Locus of arch not determinable | 3 |

respiratory tract anomalies in such patients include EA–TEF[117] and pulmonary hypoplasia[118]. Bowen and Parry's patient had the Goldenhar anomalad, a complex including ear anomalies, epibulbar ocular dermoids, vertebral anomalies and hemifacial microsomia. Lung anomalies are reported

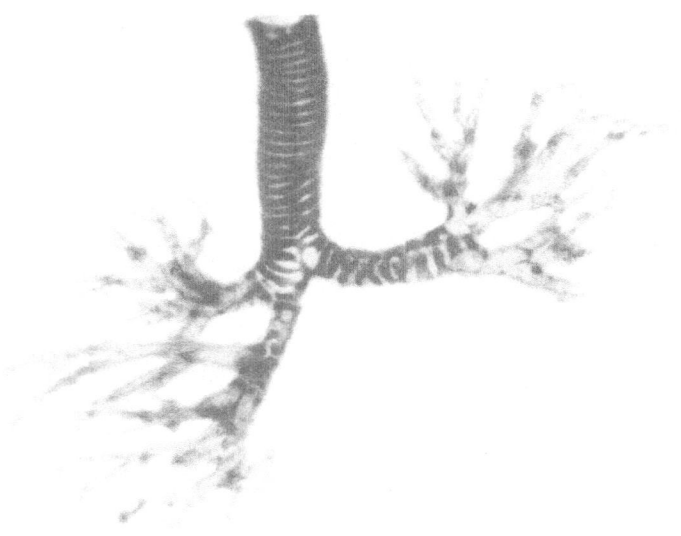

**Figure 2.5** Dissected, stained, cleared tracheobronchial tree of infant with EA–TEF, showing origin of right upper lobe bronchus at carinal level (= 'carinal trifurcation')

in only about 10 % of patients with the Goldenhar anomalad. What fraction of patients with origin of the right main bronchus from the oesophagus have this complex is uncertain, but this could be a valid association.

A possible explanation of origin of one main bronchus from the oesophagus is discussed above, where the question of whether bronchial (lung) buds may arise from the ventral pharynx before the tracheal bud separates is considered. Separation of only one of the independent lung buds from the pharyngo-oesophagus by the tracheo-oesophageal septum could explain so-called total sequestration of the right lung[119]. Why this should occur only on the left, leaving the right main bronchus attached to the oesophagus, is not apparent.

*Abnormal segmental and peripheral bronchial branching patterns*
A wide variety of variations in the patterns of these structures occurs. They are detailed for both lungs generally by Atwell[120], for the right upper lobe by le Roux[121] and by Roshe[122], and for the left upper lobe by le Roux[123]. Data on whether any of the abnormal patterns described by these workers are associated with respiratory tract or other anomalies are too poor to justify their more detailed discussion at this time. Such studies would be of interest. Possibly of most clinical importance is the variation called 'accessory cardiac bronchus', which arises from the medial aspect of the (right) bronchus intermedius. This condition shows male preponderance, but is not apparently familial, and possibly causes liability to infection[124–126]. The condition called 'left tracheal bronchus' actually is translocation of the apicoposterior

A

B

**Figure 2.6** A, B, Posterior view of partially dissected tracheobronchial tree of patient with transposition of great arteries (TGV), showing high origin of right upper lobe bronchus (at carinal level), compared to normal pattern (B)

segmental bronchus of the left upper lobe to the left main bronchus; emphysema of the area supplied is suggested to be due to pressure on the ectopic bronchus by the left pulmonary artery[127]. That right tracheal accessory or pre-eparterial bronchus is a relatively common minor anomaly is

36

well known (see reference 29); Bremer found this bronchus in 4 of 81 human embryos, explaining its frequency[128].

### Bronchial stenosis

Localized bronchial stenoses considered to represent congenital anomalies are reported (for example, affecting right main bronchus[129] and right lower lobe bronchus[130]). Multifocal stenosis of third and fourth order bronchi has been called 'bronchial coarctation'[131]. The more common phenomenon of compression of bronchi by enlarged blood vessels or cardiac chambers in patients with heart disease is reviewed by Stranger et al.[132], and Kienast et al.[133]. The left main bronchus is particularly liable to such compression, and in various disorders with heart disease, especially if pulmonary hypertension is present, it can be compressed by the left pulmonary artery, by the left atrium, or by a generally enlarged heart.

### Congenital bronchomalacia

This term is applied to diffuse, rather than local, narrowing of a bronchial segment. Most reports concern the left main bronchus (for example, references 134, 135).

Godfrey[136] described the association of pectus excavatum and bronchomalacia of the left main bronchus. He offered no specific explanation for the association, but compression by displaced heart or blood vessels would seem by Van Benthem et al.[137]. (See also Tracheomalacia, above, for discussion syndrome of mental retardation, dolichocephaly, muscular hypoplasia, reduced fatty panniculus, agenesis of the testes and chest deformities reported by Van Benthem et al.[137]. (See also Tracheomalacia, above, for discussion of tracheobronchial abnormalities, including abnormalities of calibre, in skeletal dysplasia syndromes.)

### Bronchial atresia

The best defined bronchial atresia syndrome consists of a local bronchial atresia (most commonly of the apicoposterior bronchus of the left upper lobe) with emphysema of the affected area and a radiologically visible mucus plug in the bronchus distal to the site of atresia[138]. Meng et al.[139] found that 24 of 36 reported patients with this pattern had the left upper lobe affected, 7 had the right upper lobe affected, 3 the right lower lobe and one each the right middle and left lower lobes. Although the condition is sometimes described as a form of lobar emphysema, the affected area is typically not significantly over expanded and does not cause much compression of other lung areas; most often the lesion causes no symptoms and is found 'accidentally' by X-ray. The emphysema is assumed to be the result of collateral air flow through the canals of Lambert and the pores of Kohn[140,141] but why this should occur in this particular syndrome and not in many other circumstances of bronchial obstruction cannot be explained at present. The bronchial atresia is assumed to be congenital[142]. Haller et al.[143] described a 10-year-old patient who, at age 1 day, showed radiodensity of the involved area ('retained fetal lung fluid'), but by age 6 days radiolucency. Systemic arterial supply to the affected area has been reported[144], and other reports of the rare occurrence of intralobar sequestration in the upper lobes may actually refer to this entity, but how often such systemic arterial supply occurs in this condition is not at all clear. Pressure by the (left) aortic arch or by the ductus arteriosus on the left upper lobe bronchus has been proposed to explain the predominant location of this form of bronchial atresia[145], but this seems unlikely because:

(1) the condition does not always involve the left upper lobe,

(2) bronchial compression by abnormally placed or enlarged vessels does not ordinarily cause atresia of the compressed bronchial segment(s) (see *Bronchial stenosis*, above), and

(3) as is discussed above (see *Tracheal compression by abnormal arteries in patients with the hypoplastic left heart complex*), the ductus arteriosus is not normally in close contact with the left upper lobe bronchus.

### Congenital lobar emphysema
This condition is distinguished clinically from the previous one by marked overexpansion of the affected area(s), with compression of other areas of lung, mediastinal shift, and typically earlier onset of respiratory symptoms. Bronchial compression by bronchogenic cysts or blood vessels can cause such lobar emphysema, as can bronchial mucus plug, but the most common cause is assumed to be local bronchial cartilage deficiency, with bronchial collapse and air-trapping[17,29,146]. In the review by Zatzkin et al.[147], however, in only 12 of 31 reported patients with lobar emphysema not due to bronchial compression by blood vessels was a bronchial cartilage abnormality demonstrated. The clinical picture of congenital cystic adenomatoid malformation of the lung (CCAM) can be similar; in CCAM the locus of air-trapping appears to be in the more peripheral lung tissue. An increased association of congenital heart disease, particularly patent ductus, with lobar emphysema is well recognized (e.g., references 148,149), but the association could well reflect bronchial compression by vessels rather than an association of bronchial and cardiovascular developmental errors. Dilatation ('bronchiectasis') of the left upper lobe bronchus of an infant with lobar emphysema, with the site of air-trapping not proven but assumed to be at the segmental bronchial level (and therefore multifocal), has also been described[150].

## Lungs

### The concept of the hilar lip
The general features of the development of the lungs are presented above. In their studies on the normal development of intrapulmonary bronchial branching patterns, Bucher and Reid[4] and Horsfield and Cumming[5] have noted that the zone of pulmonary parenchyma around the hilum of the lung is supplied by bronchial branches which follow a (semi) circular recurrent course from the intrasegmental bronchi, rather than from the more closely adjacent lobar bronchi. As a result, the pulmonary parenchyma around the hilum has the shortest axial bronchial pathway of any area of the lung. As Figure 2.7 shows, the lobar bronchi and main pulmonary arteries and veins occupy an

**Figure 2.7** Hilar region of lung, specimen injected with vinylite, showing the cup-shaped hilar depression occupied by the major bronchi, major pulmonary vessels, hilar lymph nodes etc., showing the 'hilar lip' of pulmonary parenchyma whose blood and air supply follows an approximately 135° retrograde course. The anatomy of the hilar lip region, and the fact that the peripheral air-passages in this region have the shortest axial distances of any area of the lungs are discussed by Horsfield and Cumming and Bucher and Reid[5,4], and is diagrammed by the former workers

area, protruding into the lung, shaped like a ball sliced off-centre, or a cup with an incurved lip, the area of parenchyma with recurrent bronchial (and vascular) supply, as described by the authors cited, being the 'hilar lip'. Whether the hilar lip shows selective liability to various acquired or diffuse pulmonary parenchymal diseases, or is preferentially affected in any broncho pulmonary malformation syndromes, is not known.

*Ectopia (herniation of the lungs)*
Herniation of the lungs upward into the neck, through the diaphragm into the abdomen, and through intercostal spaces into the subcutaneous tissue is reported in iniencephaly, and in Klippel–Feil syndrome (? = mild iniencephaly)[151,152]. That the cause of the herniation was reduced intra-thoracic space from the hyperextension of the spine seen in iniencephaly was proposed by the authors cited, but this seems improbable since reduced

intrathoracic space typically causes pulmonary hypoplasia (see below). The combined weight of the lungs in the fetus with Klippel–Feil syndrome described by Chaurasia and Singh[152] was 115 g (normal combined lung weight for crown–rump length = 20 g), suggesting that the condition actually reflects 'idiopathic pulmonary hypertrophy' (really hyperplasia). Quantitative studies, of the type done by Lynne Reid and co-workers, on the lungs in iniencephaly/brevicollis would be of value. Cervical herniation of the right lung apex, possibly through a defect in Sibson's fascia, has been reported in an infant with laryngeal hypoplasia due to the cri-du-chat (5p − ) syndrome[153], with the suggestion that the herniation was due to forceful expiration against the stenotic larynx. This proposal also seems unlikely, since laryngeal stenosis is a relatively common anomaly, and since protrusion of the lung apex through Sibson's fascia is, in infancy, actually an anatomical variant rather than a disease entity, with no apparent specific implications for other lesions in patients who show it, and requiring no treatment[154].

### Horseshoe lung
In three patients reported (see reference 155), the lungs were fused behind the heart and anterior to the oesophagus. Other congenital anomalies are typically present, but whether this reflects a specific multiple anomaly constellation (anomalad) is not clear.

### Pulmonary agenesis (aplasia)
Bilateral pulmonary agenesis is rare, and is regularly associated with other major anomalies[156,157]; whether the basic problem is in the mesoderm rather than in the endoderm (compare reference 92) is not known. Schechter[158] presents pathogenetic and pathophysiological considerations. Other congenital malformations present in patients with unilateral pulmonary agenesis include vertebral anomalies[159], a variety of patterns of congenital heart disease[160,161] (with patent ductus arteriosus associated with absence of the left lung the most frequent[162]), EA–TEF, and gastrointestinal, renal, and extremity anomalies of the sort included in the expanded anomaly complex(es) associated with EA–TEF, as discussed above. At least unilateral pulmonary agenesis appears to be a valid component of the expanded EA–TEF anomalad. Maltz and Nadas[163] reported what would appear to be a different syndrome, with unilateral pulmonary agenesis associated with a digital-radial dysplasia malformation pattern resembling (but to be differentiated from) that of the Holt–Oram syndrome[164]. A similar patient described by Frias and Felman[165] also had absence of the ipsilateral pectoralis major muscle and of the hemidiaphragm, and these authors proposed that the pattern was an extreme expression of the Poland anomaly. Although not usually showing familial occurrence, the Poland anomaly has been reported with dominant behaviour pattern, and family studies of patients with unilateral pulmonary agenesis could help clarify how many different entities can produce this lesion.

Ryland and Reid[166] found the number of bronchial generations reduced, but the number of alveoli approximately twice normal in a 3-year-old male with absence of the right lung, confirming that alveolar differentiation and multiplication does not require the normal number of proximal airway generations[2], but indicating that the remaining lung was actually not embryologically normal. They proposed that the greater than normal (for one lung) alveolar number reflected increased intrathoracic space available for lung growth *in utero*, and concluded that the alveolar multiplication beyond normal did not continue after birth.

### Lobar or segmental absence of lung tissue
This circumstance is rarely reported, presumably because of its lack of clinical significance in the absence of other anomalies (see reference 17). Storey and Marrangoni[167] reported absence of the left lower lobe bronchus with stenosis of the left main bronchus; the frequency of this association is not known.

### Pulmonary hypoplasia
The literature on pulmonary hypoplasia is extensive, and only selected aspects can be covered here. Theoretically, pulmonary hypoplasia could be due to reduced number, or reduced size, or both, of air-passages and alveoli. Reid and co-workers[2,168] have found that there is generally reduction in number of both air-passages and alveoli, so that hypoplastic lungs are, in a sense, basically 'truncated' rather than 'miniature'. Emery and Mithal[169], employing a subsequently widely used method of determining the number of alveoli in pulmonary lobules by counting the number of alveolar septal intercepts along a line running from a terminal respiratory bronchiole to the connective tissue septum at the edge of the lobule, found the number of alveoli per terminal lung unit at various fetal and postnatal ages to be as follows.

| Age | Calculated number alveoli per terminal lung unit | Increase ratio compared to term |
|---|---|---|
| Fetal weeks | | |
| 24–27 | 42 | −8 |
| 32–35 | 130 | −3 |
| 36–39 | 340 | 1 |
| | (i.e. value at birth) | |
| Postnatal months | | |
| 10–15 | 1370 | 4 |
| Years | | |
| 3 | 1556 | 5 |
| 5 | 1715 | 5+ |
| 8 | 2200 | 6 |
| 10 | 2630 | 8 |
| 12 | 3220 | 10. |

The rate of increase of alveoli per terminal lung unit with age in these data is less than the rate of increase in total lung alveoli, as would be expected from the reported postnatal increase in the number of respiratory bronchioles[7]. These values can be employed to estimate the degree of lobular growth retardation in specimens with pulmonary hypoplasia.

Swischuk *et al.*[170] classify pulmonary hypoplasia as follows (thereby specifying or implying relevant pathogenetic mechanisms):

(1) Pulmonary hypoplasia due to reduced intrathoracic space from thoracic compression, including: (a) pulmonary hypoplasia associated with renal agenesis, obstructive uropathy and other urinary tract abnormalities, the renal agenesis of sirenomelia etc. (the 'oligohydramnios tetrad')[171–175]; (b) pulmonary hypoplasia of prolonged amniotic fluid leak[176,177] and (c) Hypoplasia with elevation of the diaphragm, including abdominal masses, infantile polycystic disease of liver and kidneys, ascites and 'eventration' of the diaphragm.

(2) Pulmonary hypoplasia due to abnormality of the thoracic wall, including the skeletal dysplasias discussed above[178,179], and disorders with muscular weakness, possibly including congenital muscular dystrophy, infantile spinal muscular atrophy and congenital myasthenia. The interesting syndrome of ankyloses, facial anomalies and pulmonary hypoplasia has had much attention in the recent literature[180–186]; whether it should be placed in this category, or whether the pulmonary hypoplasia is a more fundamental component of the disease process, is not certain, although the former concept is generally accepted.

(3) Pulmonary hypoplasia due to space-occupying intrathoracic lesions, including diaphragmatic hernia, chylo/hydro-thorax (including hydrops fetalis), and intrathoracic masses such as congenital cystic adenomatoid malformation of the lung (see below), cysts and cystic hygroma[187].

(4) Primary pulmonary hypoplasia – whether prolonged amniotic fluid leak and neuromuscular diseases have been adequately excluded in patients proposed to have primary pulmonary hypoplasia, either as a sporadic or a familial occurrence, is impossible to state[188,189].

*Congenital pulmonary hypoplastic emphysema*
Henderson et al.[168] reported a 9-month-old girl whose transradiant hypoplastic left lung showed reduction in alveolar number, but alveoli with nine times normal volume. These workers summarized the categories of lobar distribution of emphysema, in terms of alveolar size and number, as follows:

(1) polyalveolar lobe, with normal number of airway branching generations and increased number of alveoli,
(2) obstructive emphysema, with normal number of bronchial generations and of alveoli,
(3) bronchial atresia (see above), with number of bronchial generations and alveolar numbers normal, and
(4) hypoplastic emphysema, with reduced numbers of bronchial generations and of alveoli.

*Scimitar syndrome*
The classical criteria of the scimitar syndrome include[190] hypoplasia of the right lung and pulmonary artery, dextroposition of the heart, systemic arterial supply to the right lower lobe by multiple small arteries traversing the diaphragm, and complete anomalous venous drainage of the right lung to the inferior vena cava above the liver (the anomalous vein curves downward

in the right lung field as seen by X-ray, with a scimitar shape, whence the name of the syndrome). The right upper lobe bronchus is typically hypoplastic or absent[4,191-3]. Familial occurrence, consistent with autosomal dominant inheritance pattern, has been reported[192]. Since possibly the most constant feature of the complex is the anomalous right pulmonary vein[193], it has been called the 'vena cava–bronchovascular syndrome'[194]. Obstruction of the abnormal vein has been suggested to be the cause of the pulmonary hypoplasia[195], but this suggestion is refuted by the absence of comparable lung changes in the relatively common syndrome of obstructed total anomalous pulmonary venous return, and by the observation that the scimitar vein may actually drain into the left atrium[196]. Congenital heart disease, of several types, may also be present in patients with the scimitar syndrome (8 of 22 patients reviewed by Jue et al.[197]); tetralogy of Fallot, patent ductus arteriosus, ventricular septal defect and pulmonic stenosis have been reported.

*Pulmonary sequestration*
The spectrum of anomalies included in the category of pulmonary (or bronchopulmonary) sequestration is broad, and the literature on the topic is large. The review by Carter[198] provides an extensive bibliography.

The definition of sequestration is a mass of pulmonary tissue which does not communicate with the tracheobronchial tree through a normally located bronchus, and is supplied by an anomalous systemic artery[199]. If the 'sequestered' pulmonary tissue is within the visceral pleura of the lung the condition is called intralobar sequestration, and if not, extralobar. The three components of the definition, and the application of one or two of them to some specimens, explain the inclusion by some workers of what appear to be different disorders, such as origin of one bronchus from the oesophagus and the scimitar syndrome, in the category of sequestration. Boyden earlier[200], and Flye and Izant[88] later have proposed, as a general explanation of sequestration, the formation of an extra tracheobronchial bud from the embryonic foregut caudal to the normal laryngotracheal groove, with earlier origin of the supernumerary bud (closer to the normal bud) giving rise to intralobar, and later origin (farther from the normal bud) to extralobar, sequestration, and with, in either circumstance, the option of subsequent involution or persistence of the oesophageal connection. This proposal has the advantages of combining both intralobar and extralobar sequestrations under one embryological mechanism. Relevant information on embryonic induction mechanisms, abnormality of which (or of response to which) could produce such accessory lung buds, are discussed above (see under *Tracheal agenesis*)[90-92].

The classical form of extralobar sequestration is the 'Rokitansky's lobe', almost always on the left (90 %), adjacent to the lower oesophagus in the posterior cardiophrenic angle, often in association with a membrane-covered defect of the diaphragm (60 %), with the arterial supply from the lower dorsal or upper abdominal aorta and the venous drainage to the hemiazygos system[201]. In contrast, the most common form of intralobar sequestration is in the posterobasal segment of the lower lobe (two thirds left), without associated diaphragmatic defect, and with the venous drainage normal (to the pulmonary veins). Arterial supply to sequestrations from the lower thoracic aorta or coeliac axis branches can be seen both as reflecting the origin of the respiratory tract as a foregut derivative[202], and as produced by persistence of early fetal vessels[2]. Blesovsky[203] gives a detailed discussion of the pathological variants of both intralobar and extralobar sequestration, emphasizing the evidence for basic commonality of the two conditions. As might be expected from the proposal that the usual mechanism of sequestration is formation of an accessory lung bud distal to the normal (see above), sequestration involving the upper lobes is uncommon; perhaps surprisingly, it is usually on the right[204], although left lingular intralobar sequestration is described[205]. Diaphragmatic 'eventration' appears, as is discussed above, to be a component of the basic condition of extralobar sequestration[206]. Other congenital malformations, interestingly, are not particularly frequent in patients with sequestration, although multiple cervical vertebral anomalies are described[207]. The frequency of this association could well be under-represented in the literature. It is important to remember that the tissue mass of extralobar sequestration may actually lie below the diaphragm[208].

## Accessory lung with duplication of intestine

Hennigar and Choy[209] described an infant with accessory lung high on the left (with pulmonary rather than systemic arterial and venous supply) plus duplication of the terminal ileum, caecum, appendix and colon. This would appear to be a different syndrome from sequestration. Entry of a left superior vena cava into the left atrium in their patient suggests the venous pattern seen in the Ivemark asplenia syndrome, and may imply a basic derangement of situs in this malformation pattern; this is supported by the origin of the accessory lung in Hennigar and Choy's patient from a left eparterial bronchus, a pattern suggestive of pulmonary branching pattern of bilateral right lung type. (We have seen pre-eparterial bronchus in the Ivemark syndrome on both right and left sides).

## Congenital cystic adenomatoid malformation of the lung (CCAM)

Stocker et al.[210] divide CCAM into three different categories.

(1) CCAM Type 1, in which the lung lesion is composed of multiple large cysts, or of a predominant cyst plus smaller cysts. This type of CCAM frequently is a large lesion, causing mediastinal shift.

(2) CCAM Type 2, with multiple evenly distributed smaller cysts not usually over 1.25 cm (0.5 inch) in diameter. Other congenital malformations are most frequent with this form of CCAM.

(3) CCAM Type 3, with the lesion a bulky firm mass containing evenly spaced small cysts.

Microscopic features of CCAM can include peripheral cartilage deficiency, abnormal columnar epithelium, which may be mucus-secreting, in the irregular large peripheral air spaces (whence the older term of congenital bronchiectasis), and the presence of striated muscle fibres in the pulmonary interstitium. Clinical distinction of Type 1 CCAM from diaphragmatic hernia can be a problem[211], and CCAM is a cause of lobar emphysema in infancy[212]. Other anomalies reported in patients with CCAM include hydranencephaly[213], and the 'prune-belly' or 'triad' syndrome[214]. Hydramnios, fetal hydrops and stillbirth are frequent with CCAM[215–218], and are possibly the result of caval compression by the enlarged affected area(s) of lung. Except for the statement that CCAM can be considered the result of one or more kinds of 'pulmonary dysplasia' no useful pathogenetic hypotheses can be made from the existing data. The condition (or at least Types 2 and 3 CCAM) appears to illustrate a defect in control of pulmonary parenchymal growth and differentiation, perhaps similar to that apparently present in the Beckwith 'hyperplastic fetal visceromegaly' syndrome.

The condition of multiple peripheral interstitial cysts of lung reported by Ives et al.[219] in two female sibs would appear to be a different disorder from CCAM. Chronic interstitial pulmonary emphysema can produce interstitial gas-filled cysts in the lungs, but the description of Ives and co-workers does not suggest this disorder either.

## Congenital absence of one pulmonary artery, origin of one pulmonary artery from the aorta, and direct communication of the right pulmonary artery with the left atrium

Technically not respiratory tract malformations, these interesting anomalies are mentioned because they imply that differentiation of the pulmonary parenchyma is not specifically dependent on blood supply via the pulmonary artery, although Hislop et al.[220] apply the term 'congenital dysplasia of the lung' to the findings in at least some patients with absence of one pulmonary artery. Ferencz[221] discusses other relations of pulmonary vascular abnormalities to lung malformations, as presented above (scimitar syndrome, sequestration etc.). Congenital bronchiectasis has been described in association with absence of the left pulmonary artery[222]; the frequency of this pattern is not clear. Tetralogy of Fallot is definitely associated with absence of one pulmonary artery; Goldsmith et al.[223] state that, of 50 such patients in the literature, 49 had absence of the left pulmonary artery and the fiftieth dextrocardia and absence of the right pulmonary artery (? situs inversus). The interesting rare anomaly of direct communication of one main pulmonary artery with the left atrium perhaps always involves the right pulmonary artery[224,225]. These conditions clearly contain information about differences in embryological processes in the two lungs. If the explanation for the genesis of sling left pulmonary artery given above[59] is correct – initial lack of the left pulmonary artery, with establishment of a collateral circulation from the right pulmonary artery via the post bronchial plexus – it is puzzling why the association of sling left pulmonary artery with tetralogy of Fallot is not more frequent, or why 'reverse sling artery', with the right pulmonary artery running behind the trachea, does not occur.

## CONCLUSIONS

Much remains to be learned about the aetiology and pathogenesis of the many interesting malformations and malformation complexes of the respiratory tract (larynx to pulmonary parenchyma). This review has attempted to point out areas of overt ignorance, of apparent conflict between current embryological dogma and the teratological facts of life, and areas where appropriate studies could well improve our understanding of basic embryological processes and control systems in the developing respiratory tract.

## Acknowledgement

This work was supported in part by the Alma Luna Wells Foundation.

## References

1 Charnock, E. L. and Doershuk, C. F. (1973). Developmental aspects of the human lung. *Pediatr. Clin. N. Am.*, **20**(2), 275

2 Reid, L. (1977). The lung: its growth and remodeling in health and disease. *Am. J. Roentgenol.*, **129**, 777

3 Cudmore, R. E., Emery, J. L. and Mithal, A. (1962). Postnatal growth of the bronchi and bronchioles. *Arch. Dis. Child.*, **37**, 481

4 Bucher, U. and Reid, L. (1961). Development of the intrasegmental bronchial tree: the pattern of branching and development at various stages of intrauterine life. *Thorax*, **16**, 207

5 Horsfield, K. and Cumming, G. (1968). Morphology of the bronchial tree in man. *J. Appl. Physiol.*, **24**, 373

6 Matsuba, K. and Thurlbeck, W. M. (1972). A morphometric study of bronchial and bronchiolar walls in children. *Am. Rev. Resp. Dis.*, **105**, 908

7 Matsuba, K. and Thurlbeck, W. M. (1971). The number and dimensions of small airways in nonemphysematous lungs. *Am. Rev. Resp. Dis.*, **104**, 516

8 Weibel, E. (1963). *Morphometry of the human lung.* (Berlin: Springer)

9 Weibel, E. R. (1979). Morphometry of the human lung. The state of the art after two decades. *Bull. Eur. Physiopathol. Resp.*, **15**, 999

10 Wailoo, M. and Emery, J. L. (1980). Structure of the membranous trachea in children. *Acta Anat.*, **106**, 254

11 Benisch, B. M., Wood, W. G., Kroeger, G. B., Breitenbach, E. E. and Cohen, J. J. (1974). Focal muscular hyperplasia of the trachea. *Arch. Otolaryngol.*, **99**, 226

12 Tos, M. (1969). Topography of tracheal mucus glands in children. *J. Laryngol. Otol.*, **83**, 1073

13 Tos, M. (1970). Anatomy of the tracheal mucus glands in man. *Arch. Otolaryngol.*, **92**, 132

14 Tos, M. (1970). Mucus glands of the trachea in children. Quantitative studies. *Anat. Anz.*, **126**, 146

15 Tos, M. (1971). Mucous glands of the trachea in man. Quantitative Studies. *Anat. Anz.*, **128**, 136

16 Sinclair-Smith, C. C., Emery, J. L., Gadsdon, D., Dinsdale, F. and Baddeley, J. (1976). Cartilage in children's lungs: a quantitative assessment using the right middle lobe. *Thorax*, **31**, 40

17 Landing, B. H. and Wells, T. R. (1973). Tracheobronchial anomalies in children. *Perspect. Pediatr. Pathol.*, **1**, 1

18 Piperno, E. (1978). Morphogenesis and histogenesis of the cartilaginous tracheal rings in the domestic fowl (Gallus domesticus). *Anat. Anz.*, **43**, 167

19 Wailoo, M. P. and Emery, J. L. (1979). The trachea in children with tracheo-oesophageal fistula. *Histopathology.*, **3**, 329

20 Liddelow, A. G. and Campbell, A. H. (1964). Widening of the membranous wall and flattening of the trachea and main bronchi. *Br. J. Dis. Chest.*, **58**, 56

21 Mackenzie, C. F., McAslan, T. C., Shin, B., Schellinger, D. and Helrich (1978). The shape of the adult human trachea. *Anesthesiology.*, **49**, 48

22 Tucker, J. A., Vidic, B., Tucker, G. F. Jr. and Stead, J. (1976). Survey of the development of laryngeal epithelium. *Ann. Otol. Rhinol. Laryngol.*, **85**, (Suppl), **30**, 1

23 Emery, J. L. and Haddadin, A. J. (1971). Squamous epithelium in respiratory tract of children with tracheo-esophageal fistula. *Arch. Dis. Child.*, **46**, 236

24 Haddadin, A. J. and Emery, J. L. (1971). Pulmonary retention simulating pneumonia as a cause of death in children with tracheoesophageal fistula. *Surgery*, **70**, 311

25 Goldman, A. S., Schochet, S. S. and Howell, J. T. (1980). The discovery of defects in respiratory cilia in the immobile cilia syndrome. *J. Pediatr.*, **96**, 244

26 McDougall, J. (1978). Endocrine-like cells in the terminal bronchioles and saccules of human fetal lung: an ultrastructural study. *Thorax*, **33**, 43

27 Reid, L. and Rubino, M. (1959). The connective tissue septa in the foetal human lung. *Thorax*, **14**, 3

28 Landing, B. H. (1957). Anomalies of the respiratory tract. *Pediatr. Clin. N. Am.* (Symposium on Respiratory Disorders, Feb. 1957), 73–102

29 Landing, B. H., with the technical assistance of Dixon, L. G. (1979). Congenital malformations and genetic disorders of the respiratory tract (larynx, trachea, bronchi, lungs). *Am. Rev. Resp. Dis.*, **120**, 151

30 Atkins, J. P. (1962). Laryngeal problems of infancy and childhood. *Pediatr. Clin. N. Am.*, **9** (4), 1125

31 Holinger, P. H. and Brown, W. T. (1967). Congenital webs, cysts, laryngoceles and other anomalies of the larynx. *Am. Otol. Rhinol. Laryngol.*, **76**, 744

32 Holinger, P. H., Johnstone, K. C. and Schild, J. A. (1962). Congenital anomalies of the tracheobronchial tree and of the esophagus. Diagnosis and treatment. *Pediatr. Clin. N. Am.*, **9**(4), 1113

33 Holinger, P. H. (1964). Congenital anomalies of the tracheobronchial tree. *Postgrad. Med.*, **36**, 454

34 Boyden, E. A. (1955). Developmental anomalies of the lungs. *Am. J. Surg.*, **89**, 79

35 Wier, J. A. (1960). Congenital anomalies of the lung. *Ann. Intern. Med.*, **52**, 330

36 Landing, B. H. (1975). Syndromes of congenital heart disease with tracheobronchial anomalies. *Am. J. Roentgenol. Rad. Ther. Nucl. Med.*, **123**, 679

37 Williams, P. L., Wendell-Smith, C. P. and Treadgold, S. (1969). *Basic Human Embryology.* 2nd Edn. (Philadelphia: Lippincott)

38 Hånallah, R. and Rosales, J. K. (1975). Laryngeal web in an infant with tracheoesophageal fistula. *Anesthesiology*, **42**, 96

39 Ward, P. H., Engel, E. and Nance, W. E. (1968). The larynx in the cri-du-chat (cat cry) syndrome. *Trans. Am. Acad. Ophthalmol. Otolaryngol.*, **72**, 90

40 Temtamy, S. A., Levin, L. S., Miller, J. D., McClanine, K. and Goldie, W. (1975). Severe Mohr syndrome or mild Majewski syndrome. *Birth Defects: Orig. Art. Ser.*, **11**(5), 342–343. (New York: National Foundation).

41 Kaufmann, H. J., Mahboubi, S., Spackman, T. J., Capitanio, M. A. and Kirkpatrick, J. (1976). Tracheal stenosis as a complication of chondrodysplasia punctata. *Ann. Radiol.*, **19**, 203

42 Goldbloom, R. B. and Dunbar, J. S. (1960). Calcification of cartilage in the trachea and larynx in infancy associated with congenital stridor. *Pediatrics*, **26**, 669

43 Baker, D. C. Jr. and Savetsky, L. (1966). Congenital partial atresia of the larynx. *Laryngoscope*, **76**, 616

44 Smith, I. I. and Bain, A. D. (1965). Congenital atresia of the larynx. A report of nine cases. *Ann. Otol. Rhinol. Laryngol.*, **74**, 338

45 Gaskill, J. R. and Bailey, B. J. (1967). Congenital posterior clefts of the larynx and trachea. *Trans. Pacific Coast Otoophthalmol. Soc.*, **51**, 259

46 Griscom, N. T. (1966). Persistent esophagotrachea. The most severe degree of laryngo-tracheo-esophageal cleft. *Am. J. Roentgenol. Rad. Ther. Nucl. Med.*, **97**, 211

47 Imbrie, J. D. and Doyle, P. J. (1969). Laryngotracheoesophageal cleft. Report of a case and review of the literature. *Laryngoscope*, **79**, 1252

48 Beazer, R., DeSa, D. J., Freeland, A. P. and Roberton, W. R. C. (1973). Laryngo-tracheoesophageal cleft. *Arch. Dis. Child.*, **48**, 912

49 Lacassie, Y. and McKusick, V. A. (1975). The G syndrome: analysis of 2 new cases. *Birth Defects: Orig. Art. Ser.*, **11**(5), 334. (New York: National Foundation)

50 Delahanty, J. E. and Cherry, J. (1969). Congenital laryngeal cleft. *Ann. Otol. Rhinol. Laryngol.*, **78**, 96

51 Mahour, G. H., Cohen, S. R. and Woolley, M. M. (1973). Laryngotracheoesophageal cleft associated with esophageal atresia and multiple tracheoesophageal fistulas in a twin. *J. Thorac. Cardiovasc. Surg.*, **65**, 223

52 Forrester, R. M. and Cohen, S. J. (1970). Esophageal atresia associated with an anorectal anomaly and probable laryngeal fissure in three siblings. *J. Pediatr. Surg.*, **5**, 674

53 Fuzesi, K. and Young, D. G. (1976). Congenital laryngotracheoesophageal cleft. *J. Pediatr. Surg.*, **11**, 933

54 Morrison, S. G., Perry, L. W. and Scott, L. P. (1968). III Congenital brevicollis (Klippel–Feil syndrome) and cardiovascular anomalies. *Am. J. Dis. Child.*, **115**, 614

55 Cantrell, J. R. and Guild, H. G. (1964). Congenital stenosis of the trachea. *Am. J. Surg.*, **108**, 297

56 Lacasse, J. E., Reilly, B. J. and Mancer, K. (1980). Segmental esophageal trachea: a potentially fatal type of tracheal stenosis. *Am. J. Roentgenol.*, **134**, 829

57 Koopot, R., Nikaidoh, H. and Idriss, F. (1975). Surgical management of anomalous left pulmonary artery causing tracheobronchial obstruction. *J. Thorac. Cardiovasc. Surg.*, **69**, 239

58 Han, B. K., Dunbar, J. S., Bove, K. and Rosenkrantz, J. G. (1980). Pulmonary vascular sling with tracheobronchial stenosis and hypoplasia of the right pulmonary artery. *Pediatr. Radiol.*, **9**, 113

59 Sade, R. S., Rosenthal, A., Fellows, K. and Castaneda, A. R. (1975). Pulmonary artery sling. *J. Thorac. Cardiovasc. Surg.*, **69**, 333

60 Cohen, S. R. and Landing, B. H. (1976). Tracheostenosis and bronchial abnormalities associated with pulmonary artery sling. *Ann. Otol. Rhinol. Laryngol.*, **85**, 1

61 Maisel, R. H., Field, M. P., Swain, R. and Spector, G. (1974). Anomalous tracheal bronchus with tracheal hypoplasia. *Arch. Otolaryng.*, **100**, 69

62 Binet, J. P., Conso, J. F., Losey, J., Narcy, P., Raynaud, E. J., Beaufils, F., Dor, C. and Bruniaux, J. (1978). Ductus arteriosus sling: report of a newly recognised anomaly and its surgical correction. *Thorax*, **33**, 72

63 Wittenborg, M. H., Gyepes, M. J. and Crocker, D. (1967). Tracheal dynamics in infants with respiratory distress, stridor and collapsing trachea. *Radiology*, **88**, 653

64 Moore, T. C. (1963). Chondroectodermal dysplasia (Ellis–van Creveld syndrome) with bronchial malformation and neonatal tension lobar emphysema. *J. Thorac. Cardiovasc. Surg.*, **46**, 1

65 Smith, D. W., Theiler, K. and Schachenmann, G. (1966). Rib-gap defect with micrognathia, malformed tracheal cartilages, and redundant skin: a new pattern of defective development. *J. Pediatr.*, **69**, 799

66 Edwards, J. E. (1979). Congenital cardiovascular causes of tracheobronchial and/or esophageal obstruction. In Tucker, B. L. and Lindesmith, G. G. (eds.) *First Clinical Conference on Congenital Heart Disease*, pp. 49–55. (New York: Grune & Stratton)

67 Gross, R. E. (1955). Arterial malformations which cause compression of the trachea or esophagus. *Circulation*, **11**, 124

68 Mustard, W. T., Trimble, A. W. and Trusler, G. A. (1962). Mediastinal vascular anomalies causing tracheal and esophageal compression and obstruction in childhood. *Can. Med. Assoc. J.*, **87**, 1301

69 Lincoln, J. C. R., Deverall, P. B., Stark, J., Aberdeen, E. and Waterston, D. J. (1969). Vascular anomalies compressing the esophagus and trachea. *Thorax*, **24**, 295

70 Fearon, B. and Shortreed, R. (1963). Tracheobronchial compression by congenital cardiovascular anomalies in children. Syndrome of apnea. *Ann. Otol. Rhinol. Laryngol.* **72**, 949

71 Wheeler, P. C., Wolff, L. J. and Stevens, E. M. (1966). Pseudovascular ring resulting from right lung agenesis, normal aortic arch, and patent ductus arteriosus. Case report. *Am. J. Roentgenol. Rad. Ther. Nucl. Med.*, **98**, 365

72 Adams, H. D., Junod, F. L., Aberdeen, E. and Johnson, J. (1972). Severe airway obstruction caused by mediastinal displacement after right pneumonectomy in a child. A case report. *J. Thorac. Cardiovasc. Surg.*, **63**, 534

73 Szarnicki, R., Maurseth, K., deLeval, M. and Stark, J. (1978). Tracheal compression by the aortic arch following right pneumonectomy in infancy. *Ann. Thorac. Surg.*, **25**, 231

74 Idriss, F. S., Nikaidoh, H., DeLeon, S. Y. and Koopot, R. (1979). Surgery for vascular anomalies causing obstruction of the trachea and esophagus. In Tucker, B. L. and

Lindesmith, G. G. (eds.) *First Clinical Conference on Congenital Heart Disease*, pp. 125–126. (New York: Grune & Stratton)

75 Macdonald, R. E. and Fearon, B. (1971). Innominate artery compression syndrome in children. *Ann. Otol. Rhinol. Laryngol.*, **80**, 535

76 Davies, M. R. Q. and Cywes, S. (1978). The flaccid trachea and tracheo-esophageal congenital anomalies. *J. Pediatr. Surg.*, **13**, 363

77 Benjamin, B., Cohen, D. and Glasson, M. (1976). Tracheomalacia in association with congenital tracheo-oesophageal fistula. *Surgery*, **79**, 504

78 Ericson, N. O. and Söderlund, S. (1969). Compression of the trachea by an anomalous innominate artery. *J. Pediatr. Surg.*, **4**, 424

79 Keibel, F. and Mall, F. P. (eds.) (1912). *Manual of Human Embryology*. vol. 2. (Philadelphia: Lippincott)

80 Hamilton, W. J. and Mossman, H. W. (1976). *Human Embryology. Prenatal Development of Form and Function.* 4th Edn. (Baltimore: Williams & Wilkins)

81 Patten, B. M. and Carlson, B. M. (1958). *Foundations of embryology.* 3rd Edn. (New York: McGraw-Hill)

82 Thomas, J. B. (1968). *Introduction to human Embryology.* (Philadelphia: Lea & Febiger)

83 Harrison, M. R., Hanson, B. A., Mahour, G. H., Takahashi, M. and Weitzman, J. J. (1977). The significance of right aortic arch in repair of esophageal atresia and tracheoesophageal fistula. *J. Pediatr. Surg.*, **12**, 861

84 Altman, R. P., Randolph, J. G. and Shearin, R. B. (1972). Tracheal agenesis: recognition and management. *J. Pediatr. Surg.*, **7**, 112

85 Hopkinson, J. M. (1972). Congenital absence of the trachea. *J. Pathol.*, **107**, 63

86 McNie, D. J. M. and Pryse-Davies, J. (1970). Tracheal agenesis. *Arch. Dis. Child.*, **45**, 143

87 Warfel, K. A. and Schulz, D. M. (1976). Agenesis of the trachea. Report of a case and review of the literature. *Arch. Pathol. Lab. Med.*, **100**, 357

88 Flye, M. W. and Izant, R. J. (1972). Extralobar pulmonary sequestration with esophageal communication and complete duplication of the colon. *Surgery*, **71**, 744

89 O'Connell, D. J. and Kelleher, J. (1979). Congenital intrathoracic bronchopulmonary foregut malformations in childhood. *J. Can. Assoc. Radiol.*, **30**, 103

90 Spooner, B. S. and Wessells, N. K. (1970). Mammalian lung development: interactions in primordium formation and bronchial morphogenesis. *J. Exp. Zool.*, **175**, 445

91 Goldin, G. V. and Wessells, N. K. (1979). Mammalian lung development: the possible role of cell proliferation in the formation of supernumerary tracheal buds and in branching morphogenesis. *J. Exp. Zool.*, **208**, 337

92 Wessells, N. K. (1970). Mammalian lung development: Interactions in formation and morphogenesis of tracheal buds. *J. Exp. Zool.*, **175**, 455

93 Dowling, E. A., Johnson, I. M., Collier, F. C. D. and Dillard, R. A. (1962). Intratracheal goiter: A clinico-pathologic review. *Ann. Surg.*, **156**, 258

94 Morgan, C. L., Grossman, H. and Leonidas, J. (1979). Roentgenographic findings in a spectrum of uncommon tracheo-esophageal anomalies. *Clin. Radio.*, **30**, 353

95 Stephens, H. B. (1970). H-type tracheoesophageal fistula complicated by esophageal stenosis. *J. Thorac. Cardiovasc. Surg.*, **59**, 325

96 Dudgeon, D. L., Morrison, C. W. and Woolley, M. M. (1972). Congenital proximal tracheoesophageal fistula. *J. Pediatr. Surg.*, **7**, 614

97 Isenberg, S. F. and Tubergen, L. B. (1979). Unilateral complete aplasia of the inner ear with associated tracheoesophageal fistula: report of a case. *Otolaryng. Head Neck Surg.*, **87**, 435

98 Barry, J. E. and Auldist, A. W. (1974). The Vater association. One end of a spectrum of anomalies. *Am. J. Dis. Child.*, **128**, 769

99 Weigel, W. and Kaufmann, H. J. (1976). The frequency and types of other congenital anomalies in association with tracheoesophageal malformations. Radiologic studies of 83 such infants. *Clin. Pediatr.*, **15**, 819

100 German, J. C., Mahour, G. H. and Woolley, M. M. (1976). Esophageal atresia and associated anomalies. *J. Pediatr. Surg.*, **11**, 299

101 Toyama, W. M. (1972). Esophageal atresia and tracheoesophageal fistula in association with bronchial and pulmonary anomalies. *J. Pediatr. Surg.*, **7**, 302

102 Hislop, A., Hey, E. and Reid, L. (1979). The lungs in congenital bilateral renal agenesis and dysplasia. *Arch. Dis. Child.*, **54**, 32

103 Atwell, J. D. and Beard, R. C. (1974). Congenital anomalies of the upper urinary tract associated with esophageal atresia and tracheoesophageal fistula. *J. Pediatr. Surg.*, **9**, 825

104 Warren, J., Evans, K. and Carter, C. O. (1979). Offspring of patients with tracheo-oesophageal fistula. *J. Med. Genet.*, **16**, 338

105 Grosser, O., Lewis, F. T. and McMarrich, J. P. (1912). The development of the digestive tract and the organs of respiration. In Keibel, F. and Mall, F. P. (eds.) *Manual of Human Embryology*. Vol. 2, pp. 291–497. (Philadelphia: Lippincott)

106 Landing, B. H., Lawrence, T.-Y. K., Payne, V. C. Jr. and Wells, T. R. (1971). Bronchial anatomy in syndromes with abnormal visceral situs, abnormal spleen and congenital heart disease. *Am. J. Cardiol.*, **28**, 456

107 Neuhauser, E. B. D., Elkin, M. and Landing, B. H. (1952). Congenital direct communication between biliary system and respiratory tract. *Am. J. Dis. Child.*, **83**, 654

108 Kohler, H. G. (1980). Association of hypoplastic left heart complex, non-lobation of lungs, polydactyly and endmaternal hydramnios in two sibs. Presented at the *26th Annual Meeting of the Paediatric Pathology Society*, Nottingham, England

109 Gonzalez-Crussi, F., Padilla, L.-M., Miller, J. K. and Grosfeld, J. L. (1976). 'Bridging bronchus': A previously undescribed air-way anomaly. *Am. J. Dis. Child.*, **130**, 1015

110 Schmitt, M., Renard, M., Grosdidier, G. and Plenat, F. (1978). Malformation trachéobron-chique associée à une artère pulmonaire gauche aberrante. *Bull. Assoc. Anat.*, **62**, 363

111 Save, S. M., Sieber, W. K. and Girdany, B. R. (1971). Congenital bronchobiliary fistula. *Surgery*, **69**, 599

112 Wagget, J., Stool, S., Bishop, H. C. and Kurtz, M. B. (1970) Congenital broncho-biliary fistula. *J. Pediatr. Surg.*, **5**, 566

113 Kalayoglu, M. and Alcay, I. (1976). Congenital bronchobiliary fistula associated with esophageal atresia and tracheoesophageal fistula. *J. Pediatr. Surg.*, **11**, 463

114 Stigol, L. C., Traversaro, J. and Trigo, E. R. (1966). Carinal trifurcation with congenital tracheobiliary fistula. *Pediatrics*, **37**, 89

115 Lewis, J. E. and Murray, R. E. (1968). Pulmonary sequestration with bronchoesophageal fistula. *J. Pediatr. Surg.*, **3**, 575

116 Bowen, A. D. III and Parry, W. H. (1980). Bronchopulmonary foregut malformation in the Goldenhar anomalad. *Am. J. Roentgenol.*, **134**, 186

117 Keeley, J. L. and Schairer, A. E. (1960). The anomalous origin of the right main bronchus from the esophagus. *Ann. Surg.*, **152**, 871

118 Thomson, N. B. Jr. and Aquino, T. (1962). Anomalous origin of the right main-stem bronchus. *Surgery*, **51**, 668

119 Jona, J. Z. and Raffensperger, J. G. (1975). Total sequestration of the right lung. *J. Thorac. Cardiovasc. Surg.*, **69**, 361

120 Atwell, S. W. (1967). Major anomalies of the tracheobronchial tree with a list of the minor anomalies. *Dis. Chest.*, **52**, 611

121 le Roux, B. T. (1962). Anatomical abnormalities of the right upper bronchus. *J. Thorac. Cardiovasc. Surg.*, **44**, 225

122 Roshe, J. (1965). Bronchovascular anomalies of the right upper lobe. An unusual case report and review of the literature. *J. Thorac. Cardiovasc. Surg.*, **50**, 86

123 le Roux, B. T. (1962). The bronchial anatomy of the left upper lobe. *J. Thorac. Cardiovasc. Surg.*, **44**, 216

124 Mangiulea, V. G. and Stingha, R. V. (1968). The accessory cardiac bronchus. Bronchologic aspect and review of the literature. *Dis. Chest.*, **54**, 433

125 Atwell, S, W. (1966). An aberrant bronchus. *Ann. Thorac. Surg.*, **2**, 438

126 Biguery, P., Denies, J. L. and de Voogd, A. (1980). La bronche cardiaque accessoire. A propos d'un cas. Revue de la litérature. *J. Radiol.*, **61**, 69

127 Rémy, J., Smith, M., Marache, P., and Nuyts, J. P. (1977). La bronche 'trachéale' gauche pathogène. Revue de la litérature à propos de 4 observations. *J. Radiol. Electrol.*, **58**, 621

128 Bremer, J. L. (1932). Accessory bronchi in embryos. Their occurrence and probable fate. *Anat. Rec.*, **54**, 361

129 Chang, N., Hertzler, J. H., Gregg, R. H., Lofti, M. W. and Brough, A. J. (1968). Congenital stenosis of the right main stem bronchus. A case report. *Pediatrics.*, **41**, 739

130 Patronas, N. J., MacMahon, H. and Variakojis, D. (1976). Bronchial web diagnosed by bronchography. *Radiology*, **121**, 526

131 Suratt, P. M., Smiddy, J. F. and Minor, G. (1980). Bronchial coarctation. *Chest*, **77**, 237
132 Stranger, P., Lucas, R. V. and Edwards, J. E. (1969). Anatomic factors causing respiratory distress in a cyanotic congenital cardiac disease. Special reference to bronchial obstruction. *Pediatrics*, **43**, 760
133 Kienast, W., Wagner, G. and Thiemann, H. (1979). Kardiovaskulär bedingte Bronchusstenose beim Ductus arteriosus persistens. *Kinderärzt Prax.*, **8**, 402
134 Chandra Mohan Gupta, T. G., Goldberg, S. J., Lewis, E. and Foukalsrud, E. W. (1968). Congenital bronchomalacia. *Am. J. Dis. Child.*, **115**, 88
135 MacMahon, H. E. and Ruggieri, J. (1969). Congenital segmental bronchomalacia. *Am. J. Dis. Child.*, **118**, 923
136 Godfrey, S. (1980). Association between pectus excavatum and segmental bronchomalacia. *J. Pediatr.*, **96**, 649
137 Van Benthem, L. H. B. M., Driessen, O., Haneveld, G. T. and Rietema, H. P. (1970). Cryptorchidism, chest deformities and other congenital anomalies in three brothers. *Arch. Dis. Child*, **45**, 143
138 Curry, T. S. III and Curry, G. C. (1966). Atresia of the bronchus to the apical-posterior segment of the left upper lobe. *Am. J. Roentgenol. Rad. Ther. Nucl. Med.*, **98**, 350
139 Meng, R. L., Jensik, R. J., Faber, L. P., Matthew, G. R. and Kittle, C. F..(1978). Bronchial atresia. *Ann. Thorac. Surg.*, **25**, 184
140 Williams, L. E., Murray, G. F. and Wilcox, B. R. (1974). Congenital atresia of the bronchus. *J. Thorac. Cardiovasc. Surg.*, **68**, 957
141 Oh, K. S., Dorst, J. P., White, J. J., Haller, J. A. Jr., Johnson, B. A. and Byrne, W. D. (1976). The syndrome of bronchial atresia or stenosis with mucocele and focal hyperinflation of the lung. *Bull. Johns. Hopkins Hosp.*, **138**
142 Montague, N. T. and Shaw, R. R. (1974). Bronchial atresia. *Ann. Thorac. Surg.*, **18**, 337
143 Haller, J. A. Jr., Tepas, J. J. III, White, J. J., Pickard, L. R. and Rabotham, J. L. (1980). The natural history of bronchial atresia. Serial observations of a case from birth to operative correction. *J. Thorac. Cardiovasc. Surg.*, **79**, 868
144 Demos, N. J. and Teresi, A. (1975). Congenital lung malformations. A unified concept and a case report. *J. Thorac. Cardiovasc. Surg.*, **70**, 260
145 Schuster, S. R., Harris, G. B. C., Williams, A., Kirkpatrick, J. and Reid, L. (1978). Bronchial atresia: a recognizable entity in the pediatric age group. *J. Pediatr. Surg.*, **13**, 682
146 Binet, J. P., Nezelof, C. and Fredet, J. (1962). Five cases of lobar tension emphysema in infancy; importance of bronchial malformation and value of post operative steroid therapy. *Dis. Chest*, **41**, 126
147 Zatzkin, H. R., Cole, P. M. and Bronsther, B. (1962). Congenital hypertrophic lobar emphysema. *Surgery*, **52**, 502
148 Buntain, W. L., Isaacs, H. Jr., Payne, V. C. Jr., Lindesmith, G. and Rosenkrantz, J. G. (1974). Lobar emphysema, cystic adenomatoid malformation, pulmonary sequestration and bronchogenic cyst in infancy and childhood: a clinical group. *J. Pediatr. Surg.*, **9**, 85
149 Strunge, P. (1972). Infantile lobar emphysema with lobar agenesis and congenital heart disease. *Acta Paediatr. Scand.*, **61**, 209
150 Shafir, R., Jaffe, R. and Kalter, Y. (1976). Bronchiectasis: a cause of infantile lobar emphysema. *J. Pediatr. Surg.*, **11**, 107
151 Chaurasia, B. D. and Wagh, K. V. (1974). Iniencephalus with ectopic lungs. *Anat. Anz.*, **136**, 447
152 Chaurasia, B. D. and Singh, M. P. (1977). Ectopic lungs in a human fetus with Klippel–Feil syndrome. *Anat. Anz.*, **142**, 205
153 Cunningham, M. D. and Peters, E. R. (1969). Cervical hernia of the lung associated with the cri-du-chat syndrome. *Am. J. Dis. Child.*, **118**, 769
154 Grunebaum, M. and Griscom, N. T. (1978). Protrusion of the lung apex through Sibson's fascia in infancy. *Thorax*, **33**, 390
155 Cipriano, P., Sweeney, L. J., Hutchins, G. M. and Rosenquist, G. C. (1975). Horseshoe lung in an infant with recurrent pulmonary infections. *Am. J. Dis. Child.*, **129**, 1343
156 Devi, B. and More, J. R. S. (1966). Total tracheopulmonary agenesis associated with asplenia, agenesis of umbilical artery and other anomalies. *Acta Paediatr. Scand.*, **55**, 107
157 DeBuse, P. J. and Morris, G. (1973). Bilateral pulmonary agenesis, esophageal atresia and the first arch syndrome. *Thorax*, **28**, 526

158 Schechter, D. C. (1968). Congenital absence or deficiency of lung tissue. The congenital subtractive bronchopneumonic malformations. *Ann. Thorac. Surg.*, **6**, 286
159 Jones, H. E. and Howells, C. H. L. (1961). Pulmonary agenesis. *Br. Med. J.*, **2**, 1187
160 Booth, J. B. and Berry, C. L. (1967). Unilateral pulmonary agenesis. *Arch. Dis. Child.*, **42**, 361
161 Rao, B., Grootman, N., Silbert, D. and Wisoff, B. G. (1976). Patent ductus arteriosus with hypoplastic lung. *Chest*, **69**, 785
162 Jimenez-Martinez, M., Pérez-Alvarez, J. J., Pérez-Treviño, C., Rubio-Alvarez, V. and DeRubens, J. (1965). Agenesis of the lung with patent ductus arteriosus treated surgically. *J. Thorac. Cardiovasc. Surg.*, **50**, 59
163 Maltz, D. L. and Nadas, A. S. (1968). Agenesis of the lung. Presentation of eight new cases and review of the literature. *Pediatrics*, **42**, 175
164 Smith, D. W. (1970). *Recognizable Patterns of Human Malformation. Genetic, Embryologic, and Clinical Aspects*, pp. 134–135. (Philadelphia, Saunders).
165 Frias, J. L. and Felman, A. H. (1974). Absence of the pectoralis major, with ipsilateral aplasia of the radius, thumb, hemidiaphragm and lung: an extreme expression of Poland anomaly? *Birth Defects: Orig. Art. Ser.*, **10**(5), 55–60. (New York: National Foundation)
166 Ryland, D. and Reid, L. (1971). Pulmonary aplasia – a quantitative analysis of the development of the single lung. *Thorax*, **26**, 602
167 Storey, C. F. and Marrangoni, A. O. (1954). Lobar agenesis of the lung. *J. Thorac. Surg.*, **28**, 536
168 Henderson, R., Hislop, A. and Reid, L. (1971). New pathological findings in emphysema of childhood: 3. unilateral congenital emphysema with hypoplasia – and compensatory emphysema of contralateral lung. *Thorax*, **26**, 195
169 Emery, J. L. and Mithal, A. (1960). The number of alveoli in the terminal respiratory unit of man during late intrauterine life and childhood. *Arch. Dis. Child.*, **35**, 544
170 Swischuk, L. E., Richardson, C. J., Nichols, M. M. and Ingman, M. J. (1979). Bilateral pulmonary hypoplasia in the neonate. *Am. J. Roentgenol.*, **133**, 1057
171 De, N. C. and Harper, J. R. (1972). Renal agenesis and pulmonary hypoplasia. *Br. Med. J.*, **3**, 676
172 Bearn, J. G. (1960). The association of sirenomelia with Potter's syndrome. *Arch. Dis. Child.*, **35**, 254
173 Bain, A. D., Beath, M. M. and Flint, W. F. (1960). Sirenomelia and monomelia with renal agenesis and amnion nodosum. *Arch. Dis. Child.*, **35**, 250
174 Fraga, J. R., Mirza, A. M. and Reichelderfer, T. E. (1973). Association of pulmonary hypoplasia, renal anomalies and Potter's facies. *Clin. Pediat.*, **12**, 150
175 Fantel, A. G. and Shepard, T. H. (1975). Potter syndrome. Nonrenal features induced by oligoamnios. *Am. J. Dis. Child.*, **129**, 1346
176 Perlman, M., Williams, J. and Hirsch, M. (1976). Neonatal pulmonary hypoplasia after prolonged leakage of amniotic fluid. *Arch. Dis. Child.*, **51**, 349
177 Thomas, I. T. and Smith, D. W. (1974). Oligohydramnios, cause of the nonrenal features of Potter's syndrome, including pulmonary hypoplasia. *J. Pediatr.*, **84**, 811
178 Hull, D. and Barnes, N. D. (1972). Children with small chests. *Arch. Dis. Child.*, **47**, 12
179 Finegold, M. J., Katzew, H., Genieser, N. B. and Becker, M. H. (1971). Lung structure in thoracic dystrophy. *Am. J. Dis. Child.*, **122**, 153
180 Say, B., Barker, N. D. and Leichtman, L. G. (1979). Ankylosis, facial anomalies and pulmonary hypoplasia syndrome. *Arch. Dis. Child.*, **133**, 1196
181 Pena, S. D. J. and Shokeir, M. H. K. (1974). Syndrome of camptodactyly, multiple ankyloses, facial anomalies and pulmonary hypoplasia: a lethal condition. *J. Pediatr.*, **85**, 373
182 Punnett, H. H., Kistenmacher, M. L., Valdes-Dapena, M. and Ellison, R. T. Jr. (1974). Syndrome of ankylosis, facial anomalies and pulmonary hypoplasia. *J. Pediatr.*, **85**, 375
183 Pena, S. D. J. and Shokeir, M. H. K. (1976). Syndrome of camptodactyly, multiple ankyloses, facial anomalies and pulmonary hypoplasia – further delineation and evidence for autosomal recessive inheritance. *Birth Defects: Orig. Art. Ser.*, **12**, 201–208. (New York: National Foundation)
184 Elias, S., Boelen, L. and Simpson, J. L. (1977). Syndromes of camptodactyly, multiple

ankylosis, facial anomalies and pulmonary hypoplasia. *Birth Defects: Orig. Art. Ser.*, **14**(6B), 243–251. (New York: National Foundation)

185 Dimmick, J. E., Berry, K., MacLeod, P. M. and Hardwick, D. F. (1977). Syndrome of ankylosis, facial anomalies and pulmonary hypoplasia. *Birth Defects: Orig. Art. Ser.*, **13**(3D), 133–138. (New York: National Foundation)

186 Mease, A. D., Yeatman, A. W., Pettet, G. and Merenstein, G. B. (1976). A syndrome of ankylosis, facial anomalies and pulmonary hypoplasia secondary to fetal neuromuscular dysfunction. *Birth Defects: Orig. Art. Ser.*, **12**, 193–200. (New York: National Foundation)

187 Csicsko, J. F. and Grosfeld, J. L. (1974). Cervicomediastinal hygroma with pulmonary hypoplasia in the newly born. *Am. J. Dis. Child.*, **128**, 557

188 Boylan, P., Home, A., Gearty, J. and O'Brien, N. G. (1977). Familial pulmonary hypoplasia. *Ir. J. Med. Sci.*, **146**, 179

189 Mendelsohn, G. and Hutchins, G. M. (1977). Primary pulmonary hypoplasia. *Am. J. Dis. Child.*, **131**, 1220

190 Valdez-Davila, O., Avila-Varguez, J., Castaneda-Zuniga, W. R., Probst, P. and Amplatz, K. (1978). A variation of scimitar syndrome. *Fortsch. Rontgenstr.*, **128**, 271

191 Bruwer, A. J. (1955). Intralobar bronchopulmonary sequestration. *Am. J. Surg.*, **89**, 1035

192 Neill, C. A., Ferencz, C., Sabiston, D. C. and Sheldon, H. (1960). The familial occurrence of hypoplastic right lung with systemic arterial supply and venous drainage: 'scimitar syndrome'. *Bull. Johns Hopkins Hosp.*, **107**, 1

193 Farnsworth, A. E. and Amkeney, J. L. (1974). The spectrum of the scimitar syndrome. *J. Thorac. Cardiovasc. Surg.*, **68**, 37

194 Kittle, C. F. and Crockett, J. E. (1962). Vena cava bronchovascular syndrome – a triad of anomalies involving the right lung: anomalous pulmonary vein, abnormal bronchi and systemic pulmonary arteries. *Ann. Surg.*, **156**, 222

195 Massami, R. A., Alwan, A. O., Hernandez, T. J., Just, H. G. and Twakkol, A. A. (1967). The scimitar syndrome. A physiologic explanation for the associated dextroposition of the heart, maldevelopment of the right lung and its artery, and for the systemic collateral supply to the lung. *J. Thorac. Cardiovasc. Surg.*, **53**, 623

196 Morgan, J. R. and Forker, A. D. (1971). Syndrome of hypoplasia of the right lung and dextroposition of the heart: 'Scimitar syndrome' with normal pulmonary venous drainage. *Circulation*, **43**, 27

197 Jue, K. L., Amplatz, K., Adams, P. Jr. and Anderson, R. C. (1966). Anomalies of great vessels associated with lung hypoplasia. The scimitar syndrome. *Am. J. Dis. Child.*, **111**, 35

198 Carter, R. (1969). Pulmonary sequestration. *Ann. Thorac. Surg.*, **7**, 68

199 Sade, R. M., Clouse, M. and Ellis, F. H. Jr. (1974). The spectrum of pulmonary sequestration. *Ann. Thorac. Surg.*, **18**, 644

200 Boyden, E. A., Bill, A. H. Jr. and Creighton, S. A. (1962). Presumptive origin of a left lower accessory lung from an esophageal diverticulum. *Surgery*, **52**, 323

201 DeParedes, C. G., Pierce, W. S., Johnson, D. G. and Waldhausen, J. A. (1970). Pulmonary sequestration in infants and children: a 20-year experience and review of the literature. *J. Pediatr. Surg.*, **5**, 136

202 Gabriele, O. F. (1970). Arterial supply to the lung via the celiac axis. *Am. J. Roentgenol.*, **109**, 522

203 Blesovsky, A. (1967). Pulmonary sequestration. A report of an unusual case and a review of the literature. *Thorax*, **22**, 351

204 Witten, D. A., Clagett, O. T., and Woolner, L. B. (1962). Intralobar bronchopulmonary sequestration involving the upper lobes. *J. Thorac. Cardiovasc. Surg.*, **43**, 523

205 Parka, W. W. (1962). Intralobar sequestration of the lingula pulmonalis. *Dis. Chest*, **41**, 378

206 Kalter, J. E., Bubis, J. J., Wolman, M. and Panzner, Y. (1962). Diaphragmatic hernia associated with accessory lung. *Dis. Chest.*, **42**, 429

207 Arcomano, J. P. and Azzoni, A. A. (1967). Intralobar pulmonary sequestration and intralobar enteric sequestration associated with vertebral anomalies. *J. Thorac. Cardiovasc. Surg.*, **53**, 470

208 Pai, S. H., Cameron, C. T. M. and Lev, R. (1971). Accessory lung presenting as juxtagastric mass. *Arch. Pathol.*, **91**, 569

209 Hennigar, G. R. and Choy, S. H. (1958). Accessory lung with persistent left superior vena cava and duplication of intestine. *J. Thorac. Cardiovasc. Surg.*, **35**, 469

210 Stocker, J. T., Drake, R. M. and Madewell, J. E. (1978). Cystic and congenital lung disease in the newborn. *Perspect. Pediatr. Pathol.*, **4**, 93

211 Monclair, J. and Schistad, G. (1974). Congenital pulmonary cysts versus a differential diagnosis in the newborn: diaphragmatic hernia. *J. Pediatr. Surg.*, **9**, 417

212 Clark, N. S., Nairn, R. C. and Sambrook Gowar, F. J. (1956). Cystic disease of the lung in the newborn treated by pneumonectomy. *Arch. Dis. Child.*, **31**, 358

213 Taber, P. and Schwartz, D. W. (1972). Cystic lung lesion in a newborn: congenital cystic adenomatoid malformation of the lung. *J. Pediatr. Surg.*, **7**, 366

214 Weber, M. L., Rivard, G. and Perreault, G. (1978). Prune belly syndrome associated with congenital cystic adenomatoid malformation of the lung. *Am. J. Dis. Child.*, **132**, 316

215 Merenstein, G. B. (1969). Congenital cystic adenomatoid malformation of the lung. Report of a case and review of the literature. *Am. J. Dis. Child.*, **118**, 772

216 Spector, R. G., Claireaux, A. E. and Williams, E. R. (1960). Congenital adenomatoid malformation of lung with pneumothorax. *Arch. Dis. Child.*, **35**, 475

217 Parodi-Hueck, L., Densler, J. F., Reed, R. C., Poulos, P. and Shulman, M. W. (1969). Congenital cystic adenomatoid malformation of the lung. *Clin. Pediatr.*, **8**, 327

218 Aslan, P. A., Korones, S. B., Richardson, R. L. and Pate, J. W. (1970). Congenital cystic adenomatoid malformation with anasarca. *J. Am. Med. Assoc.*, **212**, 622

219 Ives, E. J., Darja, M. and Geist, S. (1976). Peripheral pulmonary cystic disease in sibs. *Birth Defects: Orig. Art. Ser.*, **12**, 187–191. (New York: National Foundation)

220 Hislop, A., Sanderson, M. and Reid, L. (1973). Unilateral congenital dysplasia of lung associated with vascular anomalies. *Thorax*, **28**, 435

221 Ferencz, C. (1961). Congenital abnormalities of pulmonary vessels and their relation to malformations of the lung. *Pediatrics*, **28**, 993

222 Bogedain, W., Carpathios, J., Kalemkeris, K. and McMahon, R. J. (1962). Congenital absence of the left pulmonary artery. Association with congenital bronchiectasia and hemophilia. *J. Am. Med. Assoc.*, **182**, 247

223 Goldsmith, M., Farina, M. A. and Shaher, R. M. (1975). Tetralogy of Fallot with atresia of the left pulmonary artery. Surgical repair using a homograft aortic valve. *J. Thorac. Cardiovasc. Surg.*, **69**, 458

224 Aytac, A. and Tuncoli, T. (1967). Direct communication of a pulmonary artery with the left atrium. First reported case of all four pulmonary veins entering the communication. *J. Thorac. Cardiovasc. Surg.*, **54**, 553

225 Tuncoli, T. and Aytac, A. (1967). Direct communication between right pulmonary artery and left atrium. Report of a case and proposal of a new entity. *J. Pediatr..*, **71**, 384

# 3
# Pulmonary sequestration

## N. E. WISEMAN AND R. I. MACPHERSON

## INTRODUCTION

### Definition

Pulmonary sequestration can be defined as an aberrant mass of pulmonary tissue with a systemic arterial blood supply and no normal bronchial connection. The term 'sequestration' was introduced by Pryce[1] in 1946 and suggests a pulmonary lesion separate from the remainder of the lung both anatomically and functionally. Two major anatomical varieties are recognized – intralobar and extralobar. Both however are part of a broad spectrum of congenital anomalies, with an anomalous blood supply to normal lung at one end, and complex bronchopulmonary-foregut malformations at the other. Thus, the anatomical changes, pathogenesis, and clinical features of pulmonary sequestration can be complex and variable.

### Intralobar sequestration

This lesion consists of a mass of abnormal pulmonary tissue that shares its visceral pleural covering with the contiguous normal lung (Figure 3.5). It receives an abnormal systemic arterial blood supply from the aorta and classically has a pulmonary venous drainage to the left atrium. Communication with the tracheobronchial tree is usually absent, although an abnormal communication is occasionally found. The lesion tends to locate in the posterior basal segment of a lower lobe; 60 % on the left side. There is a low incidence of associated congenital anomalies and it is rarely found in neonates[2]. In older literature, intralobar sequestration has been called foregut cyst, congenital bronchopulmonary cyst, and dysembryopathy[3].

### Extralobar sequestration

This lesion consists of a mass of abnormal pulmonary parenchyma that is separate from the normal adjacent lung and has its own visceral pleural covering (Figure 3.7). It receives an abnormal systemic arterial blood supply from the aorta and the venous drainage classically is systemic via the azygos system to the right atrium. There is no communication with the normal tracheobronchial tree. It is usually located posteriorly, medially, and inferiorly adjacent to the costophrenic sulcus; 90 % on the left side. There is a high incidence of associated congenital anomalies and it is found as an incidental finding on neonatal autopsies. In the past, extralobar sequestration has been termed tracheal accessory lobe, Rokitansky's lobe, and supranumerary lung[3].

## Sequestration spectrum

Intralobar and extralobar sequestrations do not always follow the simplistic patterns described in the preceding paragraphs. There are many complex variations that can occur. Intralobar and extralobar sequestrations may occur simultaneously[4], or as intermediate forms with features of both[5]. Intraperitoneal[6] and intrapericardial extralobar sequestrations have been described, as have sequestrations of an entire lobe or lung[7] (Figure 3.15). In addition there are variations in the arterial blood supply and venous drainage in both intralobar and extralobar sequestration. Some have both pulmonary and systemic blood supplies or venous drainage to the right and/or left side of the heart. The anomalies of arterial blood supply and venous drainage may also occur in the absence of sequestration[1].

Complex anomalies of the diaphragm and/or gastrointestinal tract can accompany sequestration; extralobar more often than intralobar[3]. Diaphragmatic hernias are associated with as many as 58 % of extralobar sequestrations in children[8]. Anomalies of the pericardium, chest wall, and congenital heart disease, have also been reported[9]. Other rare associated anomalies include vertebral malformations[10], intestinal duplication[11,12], and Hirschsprung's disease[13].

Fistulous communications between the sequestration and the gastrointestinal tract, usually the oesophagus (Figure 3.15) have been described[14] and sequestered lobes have been found to contain hepatic or pancreatic tissue.

Intralobar and extralobar sequestrations are thus seen to be part of a broad spectrum of congenital anomalies that can be called the 'sequestration spectrum'. One of the many lesions in this spectrum is a condition that is termed 'pseudosequestration'[15]. In this anomaly a defect in the right hemidiaphragm permits normal liver to project into the right chest (Figure 3.13). This brings with it an anomalous systemic artery that perfuses the otherwise normal right lung base (Figure 3.12). The plain film and angiographic findings can be misdiagnosed as sequestration. The combination of a right diaphragmatic defect and an anomalous systemic artery can occur with hypoplasia of the right lung and a 'scimitar shaped' anomalous pulmonary vein in the so-called 'scimitar syndrome'[16]. This lesion is a further extension of the complexity and variability of the sequestration spectrum.

# PATHOGENESIS

## Congenital sequestration

It is generally believed that all extralobar and most intralobar sequestrations are of congenital origin. A number of theses concerning the embryogenesis of pulmonary sequestration have been advanced in the past; however, no one theory satisfactorily accounts for all the features within the spectrum. In 1899 Vogel[17] proposed that extralobar sequestration occurred as a result of a mechanical pinching-off of the sequestered lung by the great vessels during development. This mechanical theory is analogous to the known embryogenesis of the azygos lobe. In 1946 Pryce[1] drew attention to the aberrant arterial blood supply to sequestration and proposed a vascular traction theory. This theory maintains that the aberrant artery from the aorta is the primary anomaly and traction from this artery during development results in separation of the sequestration from the remainder of the tracheobronchial tree. It was further proposed that the often observed cystic change found in the sequestered lung was due to injury sustained during detachment.

In 1956 Smith[18] proposed that pulmonary sequestration occurred as a result of a primary developmental failure of pulmonary arterial blood supply to a portion of lung with a resultant persistence of coelomic systemic arterial blood supply. It was further suggested that high pressure within the parenchyma resulted in secondary cystic change. In 1958 Boyden[19] reported findings in two human fetuses and pointed out that pulmonary cystic change and abnormal systemic arterial blood supply to lung occur independently. In the case of pulmonary sequestration it was proposed that the occurrence of these two abnormalities was coincidental.

Cockayne and Gladstone[20] proposed a theory of development of sequestration wherein adhesions of the pulmonary bud to coelomic endothelium result in 'sequestration' of a portion of lung. Adherence of this lung tissue to the septum transversum would interfere with diaphragmatic closure and cause a diaphragmatic defect. This theory permits an explanation for the common association of extralobar sequestration with diaphragmatic hernia. The same mechanism may serve to explain 'pseudosequestration'. The developing right lung bud could adhere to the developing liver bud, interfering with closure of the right hemidiaphragm and permitting persistence of the primitive systemic circulation to the right lung base.

In 1968 Gerle and Jaretzki[14] introduced the term 'congenital bronchopulmonary foregut malformation' in the context of a unifying concept for the embryogenesis for all types of pulmonary sequestration. This concept expanded upon an earlier proposal by Eppinger and Schauenstein[21] in 1902 who stated that intralobar sequestrations develop from an additional tracheobronchial bud arising from the primitive foregut distal to the normal lung bud with subsequent caudal migration and envelopment by the normal lung. Gerle and Jaretzki[14] studied 13 cases of pulmonary sequestration in which there were patent foregut communications and proposed that both intralobar and extralobar sequestration result from the abnormal development of a caudal accessory lung bud originating from the foregut. Early accessory budding during embryogenesis results in the formulation of bronchopulmonary tissue which is enveloped by the caudal migration of the developing normal lung (intralobar sequestration). Later budding results in the formation of bronchopulmonary tissue which appears after the foregut has elongated and remains separate from the normal lung (extralobar sequestration). The absence of a foregut communication in most cases of sequestration is explained by involution of this structure. The complete spectrum of bronchopulmonary foregut malformations described by Heitoff[22] further supports a unifying concept of embryogenesis. Heitoff noted that pulmonary sequestration has been described in association with a number of foregut remnants each of which represents a different phase of involution of the foregut communication. The associated foregut remnants include (1) a patent foregut communication[22-24] (Figure 3.15), (2) a fibrous stalk between the foregut and the sequestration[25] (obliterated communication), (3) a vascular foregut attachment[19], (4) a vascular pedicle with a foregut diverticulum adjacent to the sequestration[26], (5) a fibrous stalk with a bronchial or oesophageal duplication[22], (6) a foregut diverticulum[27] and (7) a foregut cyst[11]. Each lesion in this spectrum of anomalies has been documented in patients with pulmonary sequestration.

## Acquired sequestration

As a result of recurrent lobar (segmental) lung infection a pulmonary lesion may occur which meets the criteria of intralobar sequestration and is termed 'acquired sequestration'[28]. It has been well documented that an area of chronic lung infection may acquire a significant increase in arterial blood supply[29]. This gives rise to large systemic to pulmonary anastomoses with venous drainage through the pulmonary venous system (Figure 3.9). The large systemic arteries supplying this lesion may be bronchial arteries, vessels arising below the diaphragm, or they may take origin from the intercostal or subclavian arteries[30]. Concomitant with this increase in systemic blood supply to the lung there occurs a significant decrease in pulmonary perfusion[31]. This abnormal acquired state of pulmonary perfusion is analogous to that seen with intralobar sequestration. Chronic infection may also result in bronchial obstruction with chronic atelectasis and/or consolidation. Bronchography under these circumstances will fail to demonstrate a normal bronchial communication with the affected pulmonary segment. This picture of acquired sequestration has been observed in patients with bronchiectasis[32], lung abscess, pulmonary tuberculosis, aspergillosis and cystic fibrosis. It has futher been stated that pulmonary sequestration, said to be congenital, may be complicated by the development of specific chronic lung infections such as pulmonary tuberculosis[33]. It is clearly difficult under these circumstances to differentiate between acquired and congenital sequestration. Gebauer and Mason[34] proposed that all intralobar sequestrations were acquired and occurred as a result of chronic inflammatory change beginning early in life. This controversy remains unresolved; however, in a patient with chronic lung infection known to have a normal chest roentgenogram at an earlier age and subsequently shown to develop clinical and radiological evidence of an intralobar sequestration, the lesion is no doubt acquired (Figure 3.8).

## PATHOLOGY OF SEQUESTRATION

Pathological examination of both intralobar and extralobar sequestration reveals significant abnormalities of the blood supply to the lesion and the pulmonary parenchyma which comprises the sequestration. In most instances a large single artery (Figure 3.9) supplies a sequestration; however, in as many as 15–20% of cases the vessels may be multiple[3] (Figure 3.5). Histological examination of the arteries reveals that they are elastic and thin walled, resembling the normal pulmonary artery[35]. In 75% of cases there is evidence of atherosclerotic change and occasionally aneurysmal change is present[3]. The pulmonary artery to the lesion is usually somewhat hypoplastic[36]. Veins draining the sequestration traverse the intralobar plane and in most instances of intralobar sequestration drainage occurs via the inferior pulmonary vein to the left atrium (Figure 3.3B). The venous drainage from extralobar sequestration is to the right atrium through the azygos, hemiazygos, intercostal or innominate veins and may also occur via the inferior pulmonary vein. In rare instances intralobar sequestration may have both pulmonary and systemic venous drainage[37,38].

Iwai and Shindo[25] have classified the pulmonary parenchymal abnormality observed in sequestration into four grades. In Grade 1 lesions, the sequestration consists of an enlarged and somewhat irregular bronchus containing cartilage, bronchial glands, smooth muscle and ciliated epithelium. The abnormal systemic artery is found running parallel to the abnormal bronchus. This lesion is separated from the normal parenchyma by loose connective tissue. Grade 2 lesions are composed of several interconnecting thin walled cysts which may contain fragments of cartilage (Figure 3.10). They are lined with ciliated epithelium and often are found to contain mucopurulent material. This lesion is surrounded by a thick connective tissue capsule and is separated from the normal adjacent lung. Grade 3 sequestration consists of a polycystic mass in which the cysts are intermingled with alveoli which may be atelectatic or emphysematous (Figure 3.4). The border between this lesion and the normal lung is poorly defined. In both Grade 2 and Grade 3 lesions the peripheral cysts are noted to be larger in size and the central cysts may contain cartilage. In Grade 4 sequestrations the pulmonary parenchyma appears essentially normal with some evidence of bronchial ectasia and a discrete systemic artery runs parallel to the bronchial tree.

## CLINICAL FEATURES

### Incidence

The absolute incidence of pulmonary sequestration remains unknown although the lesion has been an incidental finding in 1% of patients undergoing pulmonary resection[39]. The lesion is slightly more common in males and intralobar sequestration is six times more common than extralobar. Most extralobar sequestrations are encountered during the first decade of life, whereas intralobar sequestration has a relatively constant incidence during the first four decades[3].

## Symptoms

Many sequestrations, particularly the extralobar variety, are asymptomatic and are incidental findings on routine chest radiography, at surgery or at autopsy (Case 5, below).

The most common symptomatology is that of chronic recurrent pulmonary infection with cough, purulent sputum, fever, chest pain and respiratory distress[3] (Case 1, below). These patients may be misdiagnosed as having pneumonia, bronchiectasis, lung abscess or empyema. Occasionally infection in a sequestration may be complicated by haemoptysis[7,40], haemothorax[41] or, rarely, pneumothorax[36]. Treatment may alleviate the acute illness but recurrent infections and persistence of the radiological signs should raise the suspicion of sequestration and lead to the appropriate radiological investigations (Case 4, below). Respiratory symptoms are the rule with intralobar sequestration but the exception with the extralobar variety.

Gastrointestinal symptoms such as dysphagia, regurgitation and haematemesis can be associated with sequestration when there is a patent communication with the gastrointestinal tract[42] (Case 6, below).

## Signs

Clinical signs of pulmonary consolidation are common in patients with sequestration. Typically there will be dullness with decreased breath sounds, and in the presence of infection rhonchi will be heard. Infants may show signs of a shift of the mediastinum away from the side of a large extralobar sequestration. Both intralobar and extralobar sequestration may present with signs of high output congestive heart failure when there is a large arterial venous shunt through the lesion[9] (Cases 2, 3, below). These patients will be found to have tachycardia and a raised pulse pressure. On auscultation, a continuous or pansystolic murmur may be heard.

## RADIOLOGICAL FEATURES

Pulmonary sequestration is a diagnosis based on radiological findings. It is usually first recognized as a lower lobe density on plain chest radiographs. An extralobar sequestration appears as a well defined triangular density, usually in the left base adjacent to the mediastinum (Figure 3.6). It can be mistaken for a diaphragmatic hernia or eventration, a cyst of foregut origin or a tumour[43]. Intralobar sequestration is a less well defined lower lobe density (Figures 3.1, 3.8). If there is a bronchial communication, air may be seen within the sequestration[44]. It can be misdiagnosed as pneumonia, lung abscess or bronchiectasis[28].

Bronchography will rarely fill the lesion, as there is usually absence of a normal bronchial communication[45]. The appearance is usually that of a normally filled bronchial tree with crowding of the basal bronchi of the adjacent lung which is displaced by the sequestered mass (Figure 3.2). Occasionally, a communication between the sequestered lung and the gastrointestinal tract can be demonstrated by an upper gastrointestinal contrast study[46] (Figure 3.15).

Arteriography is the definitive diagnostic procedure[36,46]. The aberrant systemic artery can be identified by aortography[47] and arises from either the lower thoracic aorta (Figure 3.3A) or upper abdominal aorta (Figure 3.9). Selective arteriography of this vessel can be used to delineate its nature and course more clearly. In a true sequestration, the lesion should be completely perfused and opacified (Figure 3.3A). If the anomalous vessels surround but do not perfuse the mass it may be a pseudosequestration (Figure 3.12). The venous phase of the angiogram will define its venous drainage (Figure 3.3B).

Pulmonary angiography and radionuclide pulmonary perfusion scans are of limited value[48]. Both will demonstrate decreased perfusion in the area of the pulmonary lesion. A liver–spleen scan is indicated if a pseudosequestration is suspected (Figure 3.13).

If cardiac catheterization studies are carried out using dye dilution techniques it is possible to quantitate the magnitude of the shunt through the lesion. Cardiac catheterization may also detect the presence of an associated congenital cardiac anomaly[9].


## TREATMENT

The treatment of symptomatic pulmonary sequestration has as its main objective the prevention of recurrent pulmonary sepsis. This has been most successfully achieved with surgical resection of the lesion. Ideally, surgical resection should be carried out at a time when acute infection has been controlled with the use of appropriate antibiotic therapy and the disease is quiescent. Because intralobar sequestrations usually involve one segment of lung it has been advocated by some that segmentectomy or basal segmentectomy be employed[9,49,50]. It has, however, been the experience that due to the chronicity of infection, and the ill-defined intersegmental planes which occur with chronic infection the safest surgical procedure is lobectomy[51]. In the management of extralobar sequestration it has usually been possible to carry out a simple resection of the lesion preserving normal lung[3].

A number of authors have pointed out that during the conduct of lobectomy in the management of pulmonary sequestration careful attention must be paid to the aberrant systemic arteries. These vessels have been reported to result in lethal haemorrhage and thus it is essential that they be carefully identified and divided[52]. In patients with a known communication between the sequestration and the oesophagus or stomach it is also important to divide this abnormal communicating structure. With careful surgical technique one anticipates successful treatment in all patients with pulmonary sequestration.

## SUMMARY

Pulmonary sequestration is defined as an aberrant mass of pulmonary tissue with a systemic arterial blood supply and no normal bronchial communication. Two distinct entities – intralobar and extralobar sequestration – are well recognized and appear to be only a part of the 'sequestration spectrum'. The lesion is usually of congenital origin, having its embryogenesis linked with the complex of bronchopulmonary foregut malformations. Acquired sequestrations are also recognized. The clinical presentation is variable and may include signs and symptoms referrable to the pulmonary, gastrointestinal or cardiovascular systems. The correct diagnosis is established by appropriate radiological investigation, the hallmark of which is angiography. Surgical resection is the treatment of choice and a successful outcome is usual.

## CASE HISTORIES

### Case 1 (A408983)

This infant male was noted to be tachypnoeac with grunting respirations in the neonatal period. Chest X-ray revealed a left lower lobe infiltrate. Following medical treatment for pneumonia clinical improvement was noted; however, the radiographic appearance remained unchanged (Figure 3.1). Over the first 10 months of life the infant required hospitalization on five occasions for treatment of recurrent left lower lobe pneumonia. Bronchography at 3

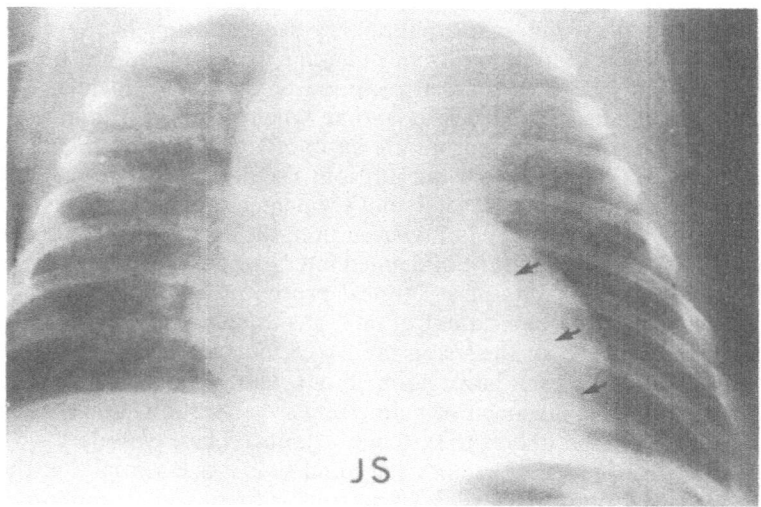

**Figure 3.1** Intralobar sequestration. Chest radiograph showing a triangular density (arrows) in left lower lobe behind the heart

months of age demonstrated filling of the left bronchial tree with displacement of the lower lobe by a mass lesion in the posteromedial aspect of the left hemithorax (Figure 3.2). At angiography an intrabolar sequestration was demonstrated with arterial blood supply via a large vessel arising from the aorta below the diaphragm (Figure 3.3). Venous drainage was through the inferior pulmonary vein to the left atrium. At 13 months of age, left lower lobectomy was performed. Pathological examination of the affected lobe

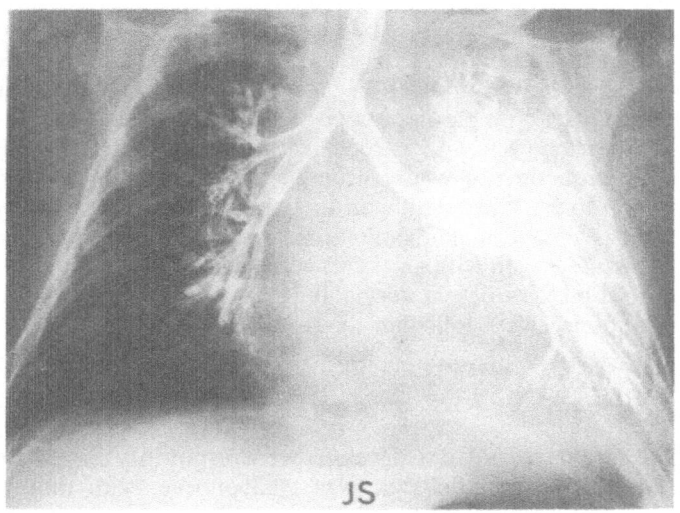

**Figure 3.2** Intralobar sequestration. Bronchogram showing normal left bronchial tree displaced by mass in left base

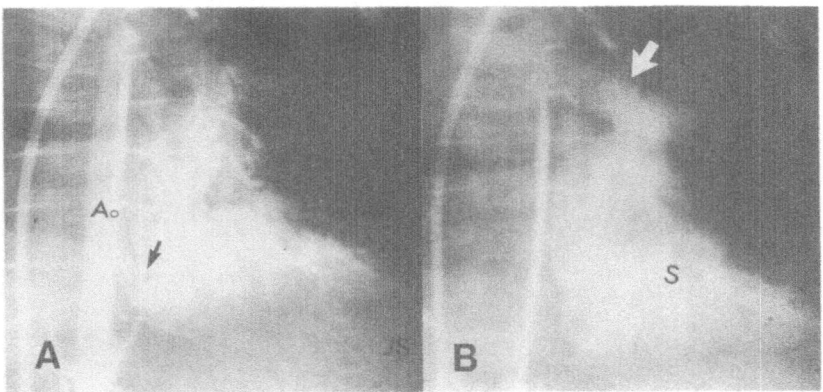

**Figure 3.3** Intralobar sequestration. Aortogram. A, arterial phase showing aberrant artery (arrow) arising from aorta (Ao) and perfusing the mass; B, venous phase showing opacified sequestration (S) draining via pulmonary vein (arrow) to left atrium

63

revealed an intralobar sequestration of the multicystic type with indistinct margins between the sequestration and the remainder of the lower lobe (Figure 3.4). The child experienced an uneventful postoperative recovery and was well at 6 years of age.

## Case 2 (554369)

This 14-month-old male presented with a 1 year history of a persistent left lower lobe infiltrate. On examination the child was noted to have bounding pulses and a continuous murmur could be heard over the left posterior chest. At angiography, an intralobar sequestration involving the posterior basal segment of the left lower lobe was diagnosed. The arterial blood supply arose from the aorta just above the diaphragm and venous drainage was through the inferior pulmonary vein. At cardiac catheterization the systemic shunt through the sequestration was calculated to be 20% of cardiac output. Following left lower lobectomy with resection of the sequestration, cardiac output returned to normal. Examination of the resected lung revealed an intralobar sequestration with an extensive systemic blood supply to both the sequestration and the adjacent normal lung (Figure 3.5). The child is well and free of symptoms 1 year following surgery.

## Case 3 (583962)

This female infant was noted to develop severe respiratory distress on the first day of life. On examination she was tachypnoeac, with indrawing and decreased air entry over the left hemithorax. Treatment for congestive heart

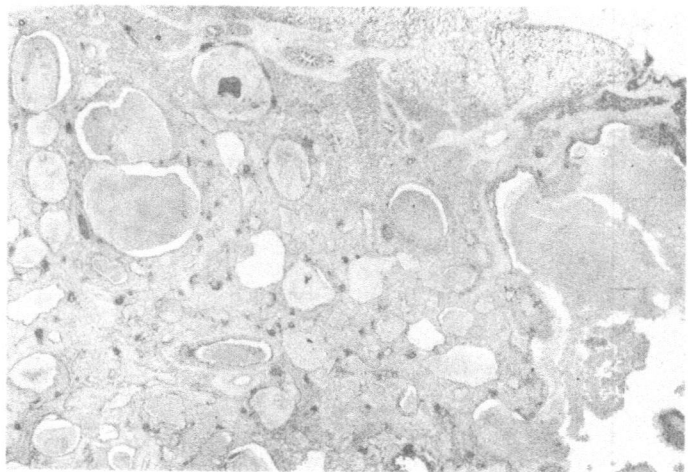

**Figure 3.4** Intralobar sequestration. Photomicrograph ( × 5) illustrating multiple bronchiolar cysts intermingled with alveolar parenchyma.

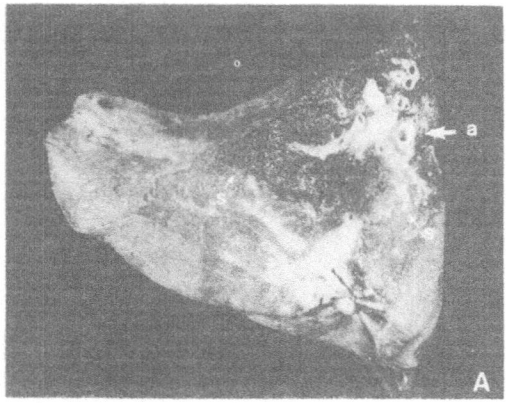

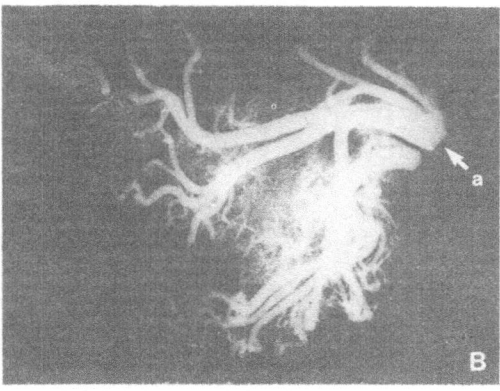

**Figure 3.5** Intralobar sequestration. A, gross photograph of left lower lobe illustrating abnormal systemic arteries (a) and margins of sequestration (s); B, radiograph of surgical specimen following perfusion of abnormal systemic arteries (a)

failure was commenced and she was placed on a ventilator. A chest X-ray on the first day of life revealed a mass in the left hemithorax (Figure 3.6). Aortography was subsequently performed and this demonstrated an extralobar sequestration receiving arterial blood from a large vessel arising below the diaphragm. At surgery the sequestration was found to be completely separate from the normal left lung with no bronchial communication (Figure 3.7). Postoperatively this infant was weaned successfully from the ventilator and made a satisfactory recovery.

## Case 4 (45374)

This boy initially presented at age 2 years with bilateral pneumonia which improved following a course of medical therapy. Over the following 13-year

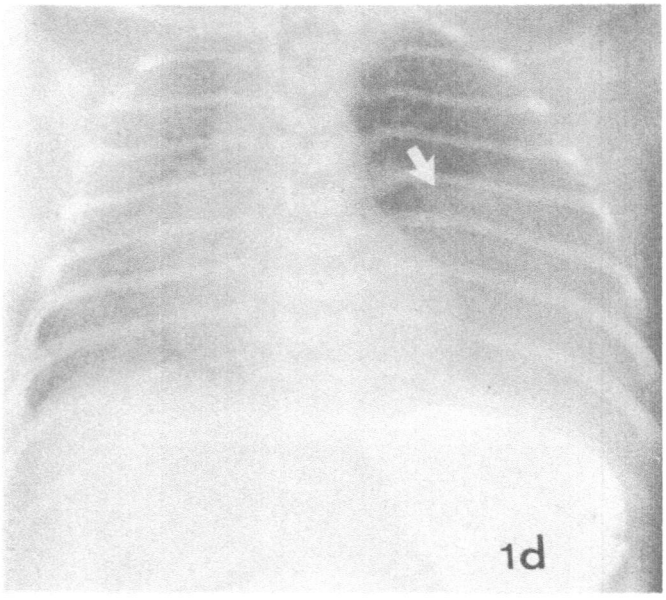

**Figure 3.6** Extralobar sequestration. Chest film in newborn showing mass in left base (arrow). Barium in stomach below diaphragm rules out diaphragmatic hernia or eventration

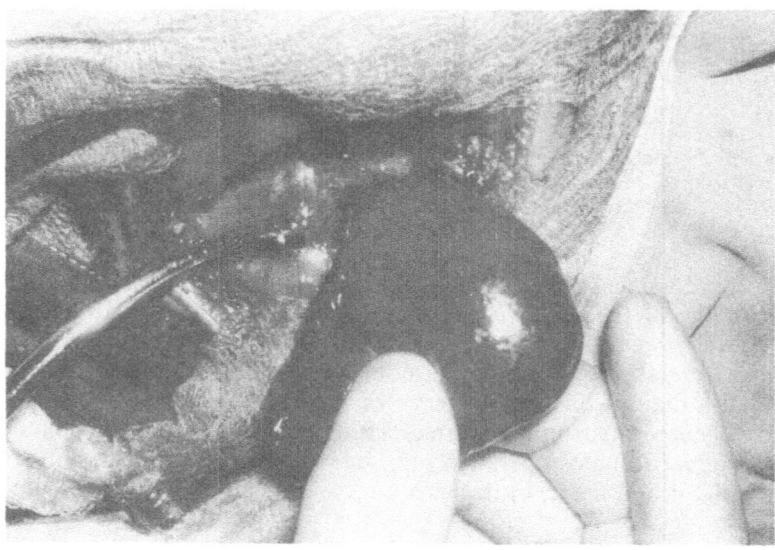

**Figure 3.7** Extralobar sequestration. Surgical photograph showing lesion separate from left lung with vascular pedicle arising below diaphragm

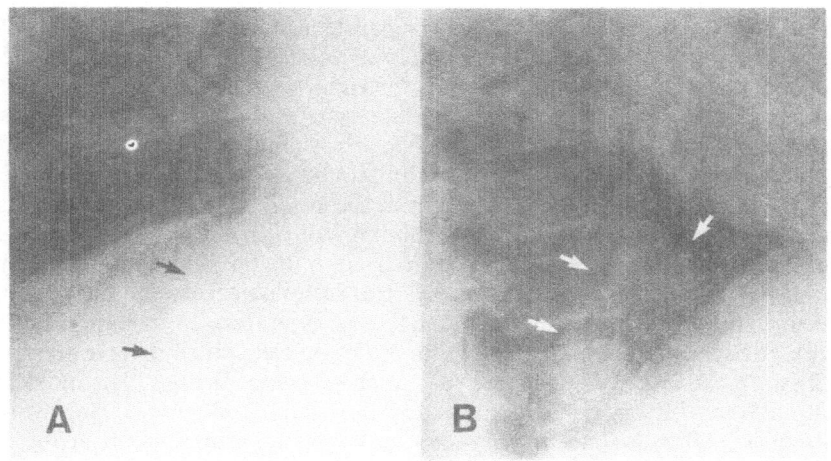

**Figure 3.8**  Acquired sequestration. Chest film. Frontal (A) and lateral (B) projections showing irregular density in right base (arrow) that is barely perceptible on frontal projection

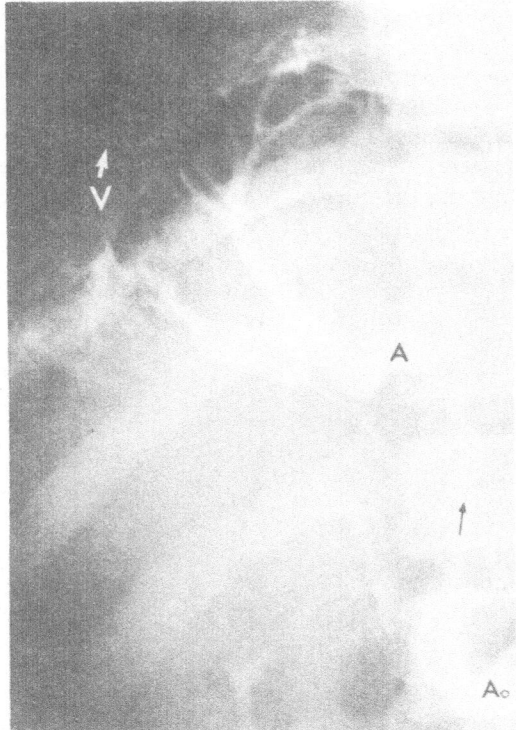

**Figure 3.9**  Acquired sequestration. Aortogram showing anomalous artery (A) arising from abdominal aorta (Ao) entering mass in right base. The venous drainage (V) is to the left atrium

67

period he was hospitalized on eight separate occasions for treatment of recurrent pneumonia involving the right lower lobe. The chest X-ray at age 15 revealed a mass lesion in the region of the right lower lobe (Figure 3.8). At angiography a large systemic artery supplying an intralobar sequestration was seen arising from the region of the coeliac artery (Figure 3.9). At surgery an intralobar sequestration was diagnosed and found to have four large systemic arteries passing through the diaphragm to the lesion. The sequestration was noted to be adherent to both the diaphragm and the oesophagus. Venous drainage from the sequestration occurred via both the inferior pulmonary venous system and the azygos venous system. Pathological examination of the specimen revealed large cystic cavities lined by respiratory epithelium (Figure 3.10). The systemic elastic vessels within the lesion were noted to have marked intimal thickening and evidence of atherosclerotic change. Histological examination of the parenchyma revealed chronic inflammatory change with pulmonary fibrosis. This lesion was considered to represent an acquired intralobar sequestration.

## Case 5 (528103)

This asymptomatic 3½-year-old girl presented with the incidental finding on chest X-ray of a soft tissue mass in the right base located posteromedially (Figure 3.11). On physical examination there was slight decrease in air entry over the right lower lobe. An aortogram and selective arterial injection revealed an aberrant artery arising just below the coeliac axis ascending to surround the pseudosequestration and supply the adjacent right lower lobe (Figure 3.12). Venography demonstrated a large abnormal venous tributary from the mass entering the inferior vena cava. A nucleide scan of the liver

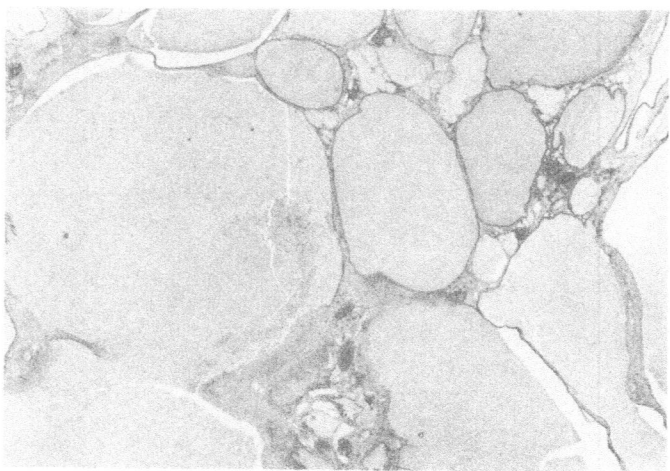

**Figure 3.10** Acquired sequestration. Photomicrograph ( × 6.3) demonstrating large cystic spaces lined by columnar respiratory epithelium and filled with proteinaceous material

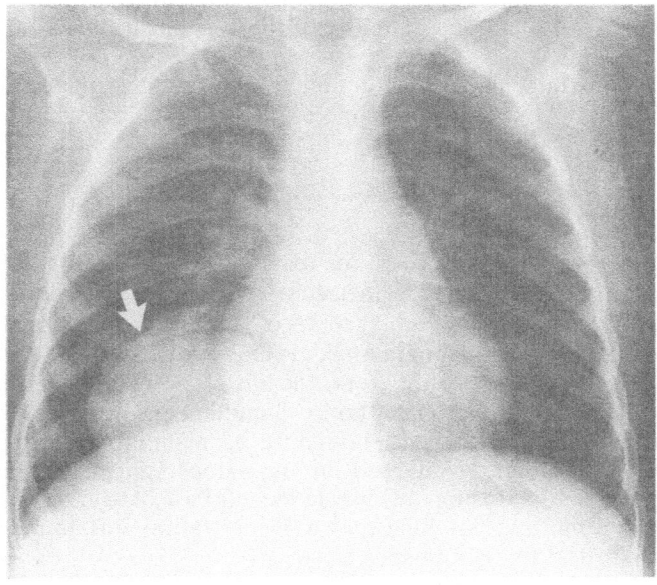

**Figure 3.11** Pseudosequestration. Chest film showing rounded mass in right base (arrow)

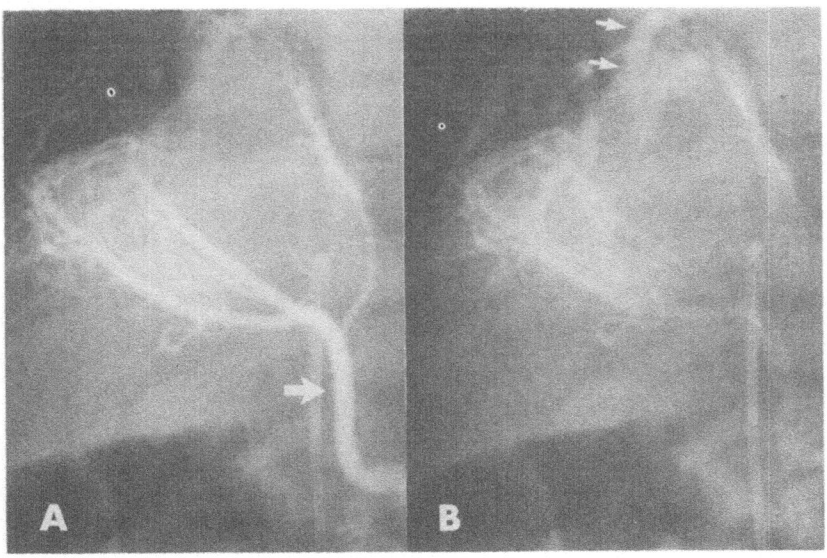

**Figure 3.12** Pseudosequestration. Angiogram. A, arterial phase showing anomalous artery (arrow) arising from aorta surrounding mass in right base but not perfusing it; B, venous phase showing venous drainage (arrows) to left atrium

confirmed the diagnosis of pseudosequestration with demonstration of herniation of a portion of the right lobe of the liver through a diaphragmatic hernia (Figure 3.13).

This case illustrates the incidental finding of a pseudosequestration in an otherwise healthy child and it was elected not to perform surgery for this lesion.

### Case 6 (BGV)

This 5-month-old Mexican American female presented with a history of respiratory distress associated with feedings. A chest radiograph revealed radiological signs consistent with agenesis of the left lung (Figure 3.14) and this was confirmed by pulmonary angiography. An oesophagram showed a bronchus-like structure arising from the distal oesophagus entering an imperceptible mass in the left hemithorax. Simultaneous with this a broncho-gram filled only the right lung and showed the left main bronchus to be absent (Figure 3.15). Aortography showed an anomalous systemic artery arising from the aorta and perfusing the mass in the left base. At surgery, there was absence of the normal left lung and a sequestration was found with its bronchus joined to the oesophagus.

This can be considered either agenesis of the left lung associated with a

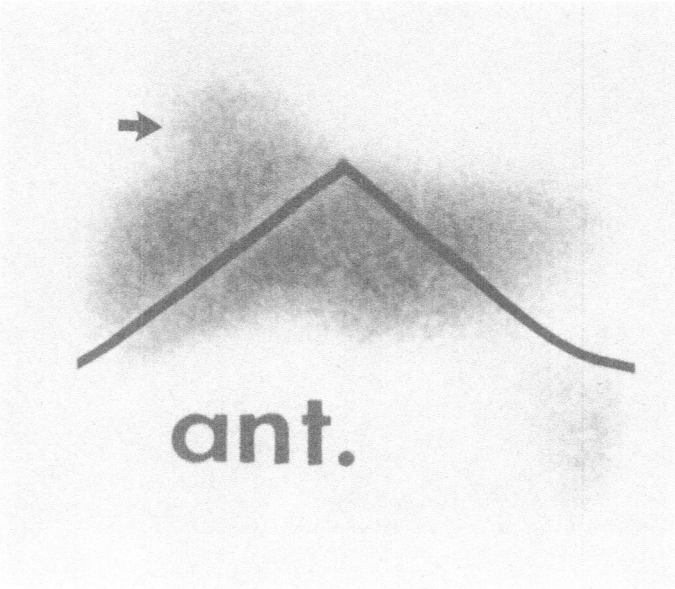

**Figure 3.13** Pseudosequestration. Radionuclide liver–spleen scan showing mass of liver tissue (arrow) projecting above right hemi-diaphragm

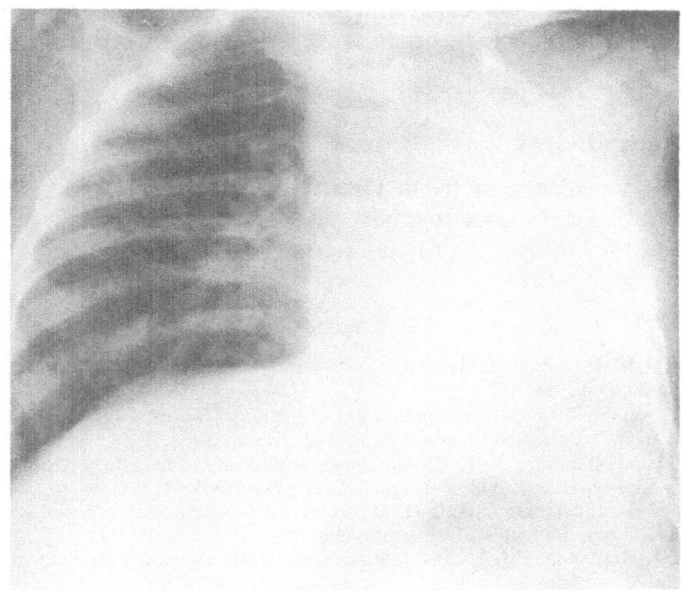

**Figure 3.14** Agenesis of left lung associated with sequestration. Chest film showing opaque left hemithorax with heart shifted to left, consistent with agenesis of left lung

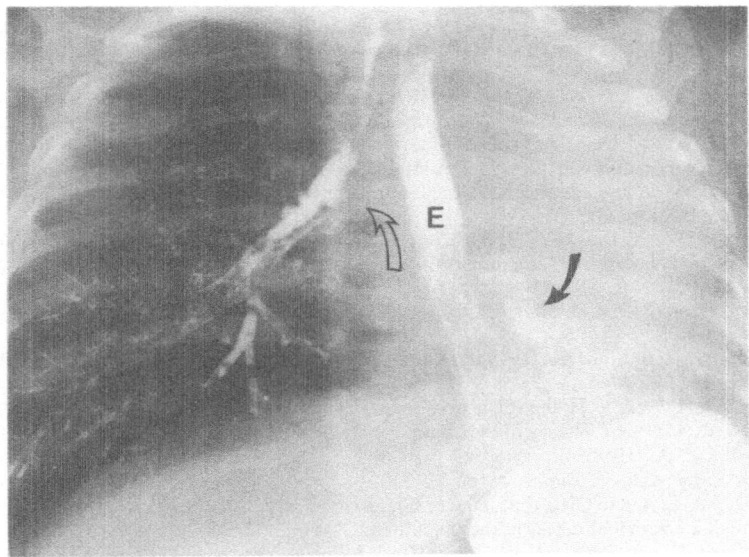

**Figure 3.15** Agenesis of left lung associated with sequestration. Simultaneous bronchogram and oesophagram showing the absence of a left main stem bronchus at the carina (open arrow). A bronchus (closed arrow) arises from the lower oesophagus (E) and enters a sequestered lobe in the left base

sequestration that communicated with the oesophagus or sequestration of the entire left lung with an ectopic origin of the left main stem bronchus from the oesophagus.

## Acknowledgements

The authors would like to thank Dr J. Hoogstraten for his assistance in preparation of the photomicrographs and Mrs P. Ryan for her secretarial assistance in preparation of this manuscript.

## References

1 Pryce, D. M. (1946). Lower accessory pulmonary artery with intralobar sequestration of lung. *J. Pathol. Bacteriol.*, **58**, 457

2 Smith, R. A. (1955). Intralobar sequestration of the lung. *Thorax*, **10**, 142

3 Carter, R. (1969). Pulmonary sequestration. *Ann. Thorac. Surg.*, **7**, 68

4 Pendse, P. and Alexander, J. (1972). Pulmonary sequestration; coexisting classic intralobar and extralobar types in a child. *J. Thorac. Cardiovasc. Surg.*, **64**, 127

5 Zumbro, G. L. and Treasure, R. L. (1975). Pulmonary sequestration; a broad spectrum of bronchopulmonary foregut malformations. *Ann. Thorac. Surg.*, **20**, 161

6 Valle, A. R. and White, M. L. (1947). Subdiaphragmatic aberrant pulmonary tissue. *Dis. Chest*, **13**, 63

7 Jona, J. Z. and Raffensberger, J. G. (1975). Total sequestration of the right lung. *J. Thorac. Cardiovasc. Surg.*, **69**, 361

8 DeParedes, C. G. and Pierce, W. S. (1970). Pulmonary sequestration in infants and children. *J. Pediatr. Surg.*, **5**, 136

9 White, J. J. and Donahoo, J. S. (1974). Cardiovascular and respiratory manifestations of pulmonary sequestration in childhood. *Ann. Thorac. Surg.*, **18**, 286

10 Arcomano, J. P. and Azzoni, A. A. (1967). Intralobar pulmonary sequestration and intralobar enteric sequestration associated with vertebral anomalies. *J. Thorac. Cardiovasc. Surg.*, **53**, 470

11 Mahour, G. H. and Wooley, M. M. (1971). Association of pulmonary sequestration and duplication of the stomach. *Int. Surg.*, **56**, 224

12 Flye, M. W. and Izant, R. J. (1972). Extralobar pulmonary sequestration with esophageal communication and complete duplication of the colon. *Surgery*, **71**, 744

13 Claman, M. A. and Ehrenhaft, J. L. (1960). Bronchopulmonary sequestration. *J. Thorac. Cardiovasc. Surg.*, **39**, 531

14 Gerle, R. D. and Jaretzki, A. (1968). Congenital bronchopulmonary–foregut malformations; pulmonary sequestration communicating with the gastrointestinal tract. *N. Engl. J. Med.*, **278**, 1413

15 Macpherson, R. I. and Whytehead, L. (1977). Pseudosequestration. *J. Can. Assoc. Radiol.*, **28**, 17

16 Roehm, J. O. and June, K. L. (1966). Radiographic features of the scimitar syndrome. *Am. J. Roentgenol.*, **86**, 856

17 Vogel, R. (1899). Zwei Fälle von abdominalem Lungengewebe. *Arch. Pathol. Anat.*, **155**, 235

18 Smith, R. A. (1956). Theory of the origin of intralobar sequestration of lung. *Thorax*, **11**, 10

19 Boyden, E. A. (1958). Bronchogenic cysts and the theory of intralobar sequestration: new embryologic data. *J. Thorac. Surg.*, **35**, 605

20 Cockayne, E. A. and Gladstone, R. J. (1917). A case of accessory lungs associated with hernia through a congenital defect in the diaphragm. *J. Anat.*, **52**, 64

21 Eppinger, H. and Schauenstein, W. (1902). Krankheiten der Lungen. *Ergeb. Allg. Pathol.*, **8**, 267

22 Heithoff, B. K. and Shashikant, M. (1976). Bronchopulmonary foregut malformations. A unifying etiological concept. *Am. J. Roentgenol.*, **126**, 46

23 Lewis, E. J. and Murray, R. E. (1968). Pulmonary sequestration with bronchoesophageal fistula. *J. Pediatr. Surg.*, **3**, 575

24  Moscarella, A. A. and Wylie, R. H. (1968). Congenital communication between esophagus and isolated ectopic pulmonary tissue. *J. Thorac. Cardiovasc. Surg.*, **55**, 672

25  Iwai, K. and Shindo, G. (1973). Intralobar pulmonary sequestration, with special reference to developmental pathology. *Am. Rev. Resp. Dis.*, **107**, 911

26  Bates, M. (1968). Total unilateral pulmonary sequestration. *Thorax*, **23**, 311

27  Boyden, E. A. and Bill, A. H. (1962). Presumptive origin of a left lower accessory lung from an esophageal diverticulum. *Surgery*, **52**, 323

28  Choplin, R. H. and Siegel, M. J. (1980). Pulmonary sequestration: six unusual presentations. *Am. J. Roentgenol.*, **134**, 695

29  Wiseman, N. E. and Reed, M. H. (1980). Broncho pulmonary arterial malformation occurring in aspergillus lung infection complicating chronic granulomatous disease. *J. Pediatr. Surg.*, (In press)

30  Pinet, F. and Froment, J. C. (1978). Angiography of the thoracic systemic arteries. *Radiol. Clin. N. Am.*, **16**, 441

31  Liebow, A. A. and Hales, M. R. (1950). Studies on the lung after ligation of the pulmonary artery. Anatomical changes. *Am. J. Pathol.*, **26**, 177

32  Victor, S. and Lakshmikanthan, C. (1972). Continuous murmur as a sequel of augmented collateral circulation in suppurative lung disease. *Chest*, **62**, 504

33  Borrie, J. and Lichter, I. (1963). Intralobar pulmonary sequestration. *Br. J. Surg.*, **50**, 623

34  Gebauer, P. W. and Mason, C. B. (1959). Intralobar pulmonary sequestration associated with anomalous pulmonary vessels; a nonentity. *Chest*, **30**, 282

35  Bruwer, A. J. and Clagett, O. T. (1954). Intralobar bronchopulmonary sequestration. *Am. J. Roentgenol. Radium Ther. Nucl. Med.*, **71**, 751

36  Mannix, E. P. and Haight, C. (1955). Anomalous pulmonary arteries and cystic disease of the lung. *Medicine*, **34**, 193

37  Turk, L. N. and Lindskog, G. E. (1961). The importance of angiographic diagnosis in intralobar pulmonary sequestration. *J. Thorac. Cardiovasc. Surg.*, **41**, 299

38  Sade, R. M. and Clouse, M. (1974). The spectrum of pulmonary sequestration. *Ann. Thorac. Surg.*, **18**, 644

39  Seitter, G. and Larson, A. (1974). Pulmonary sequestration. *Milit. Med.*, **139**, 899

40  Reichert, J. R. and Winkler, S. S. (1974). Spontaneous hemorrhage into an extralobar bronchopulmonary sequestration. *Radiology*, **110**, 359

41  Zumbro, G. L. and Green, D. C. (1974). Pulmonary sequestration with spontaneous intrapleural hemorrhage. *J. Thorac. Cardiovasc. Surg.*, **68**, 673

42  Holstein, P. and Hjelms, E. (1973). Bronchopulmonary sequestration *J. Thorac. Cardiovasc. Surg.*, **65**, 462

43  Buntain, W. L. and Isaacs, H. (1974). Lobar emphysema, cystic adenomatoid malformation, pulmonary sequestration, and bronchogenic cysts in infancy and childhood: A clinical group. *J. Pediatr. Surg.*, **9**, 85

44  Felson, B. (1972). The many faces of pulmonary sequestration. *Semin. Roentgenol.*, **7**, 3

45  Zelefsky, M. N. and Janis, M. (1971). Intralobar bronchopulmonary sequestration with bronchial communication. *Chest*, **59**, 266

46  Halasz, N. A. and Lindskog, G. E. (1962). Esophagobronchial fistula and bronchopulmonary sequestration; report of a case and review of the literature. *Ann. Surg.*, **155**, 215

47  Kenney, L. J. and Eyler, W. R. (1956). Preoperative Diagnosis of Sequestration of the Lung by Aortography. *JAMA*, **160**, 1464

48  Khalil, K. G. and Kilman, J. W. (1975). Pulmonary Sequestration. *J. Thorac. Cardiovasc. Surg.*, **70**, 928

49  Howard, R. (1963). Conservative excision in intralobar sequestration of the lung. *Lancet*, **2**; 1295

50  LeRoux, B. T. (1962). Intralobar pulmonary sequestration. *Thorax*, **17**, 77

51  Buntain, W. L. and Wooley, M. M. (1977). Pulmonary sequestration in children; a 25 year experience. *Surgery*, **81**, 413

52  Douglass, R. (1948). Anomalous pulmonary vessels. *J. Thorac. Surg.* **17**, 712

# 4
# Bronchogenic cyst

M. L. RAMENOFSKY

## INTRODUCTION

### Definition

A bronchogenic cyst is a duplication of the primitive pulmonary primordium containing all of the histological characteristics of a bronchus. A duplication arising from the foregut prior to the development of the pulmonary primordium, or arising above or below the level from which this primordium develops, cannot be classified as a bronchogenic cyst, even though it may be lined with respiratory epithelium[1-3].

### Historical aspects

The earliest reference to congenital cystic disease of the lung is that of Bartholinus in 1687[4]. Bronchogenic cyst was first reported in 1911 by Blackater[5]. The first successful surgical excision of a bronchogenic cyst was reported by Maier in 1948[6]. Maier's classification of bronchogenic cysts by their location is the most commonly used classification today.

### Occurrence

*Incidence*
Bronchogenic cysts were, at one time, considered to be extremely rare. However, with improved methods of diagnosis and the utilization of modern surgical methods, it is known that bronchogenic cysts comprise 10% of all intrathoracic masses in infancy and childhood[7-11].

*Sex and age*
There is a slight male sex preponderance in patients with bronchogenic cyst (57%:43%)[6-9,11-16]. The age at diagnosis ranges from the newborn period through the sixth decade[17]. However, the majority of bronchogenic cysts are diagnosed in patients below the age of 10 years, one third diagnosed by age 2, 10% in the neonatal period[6-8,12-15,17-20].

# EMBRYOLOGY

## Normal tracheopulmonary development

The primitive foregut gives rise to the ventral pulmonary primordium and the dorsal oesophagus[1-3]. Towards the end of the third week of development the primordium of the respiratory tract appears as an outgrowth from the ventral wall of the primitive foregut. This outgrowth is of endodermal origin. The respiratory diverticulum soon becomes separated from the foregut by the oesophagotracheal septum except at the laryngeal orifice (Figure 4.1).

After separation from the primitive foregut the respiratory primordium begins to grow in a caudal direction forming a midline tubular structure, the trachea with two lateral outpouchings, the lung buds. The lung buds grow in a caudolateral direction penetrating the splanchnic mesoderm of the coelomic cavity, known as the pleuroperitoneal canals. Once the lung buds expand to fill these canals, the primitive pleural cavities are formed. The splanchnic mesoderm is vital to bronchopulmonary development as it differentiates into cartilage, muscle, and blood vessels, whereas the endoderm provides the respiratory epithelium and the mucous glands.

Bronchial division continues and by the end of the sixth month there are approximately 17 generations of the bronchial tree. An additional six divisions of the bronchial tree develop in postnatal life.

By the seventh month the distal ends of the terminal bronchioli expand into alveoli, the lining provided by endoderm and the capillaries by mesoderm.

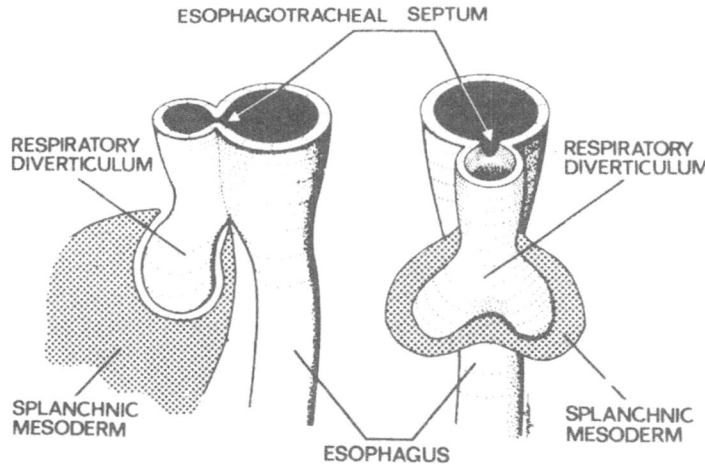

**Figure 4.1** Diagrammatic representation of the respiratory diverticulum developing as an outgrowth from the ventral wall of the primitive foregut. (Adapted from several sources)

## Embryology of cysts of tracheobronchial origin

Cysts of tracheal or bronchial origin are duplications which result from abnormal budding of the primitive pulmonary primordium, the primitive trachea, or the lung buds[2,3] (Figure 4.2). Both primitive endoderm and mesoderm must be present during the budding process so that all the characteristic components of a bronchus will be present[3]. (Although duplications of the oesophagus may be lined by respiratory epithelium, ciliated columnar, they are not considered here.) The location of these tracheobronchial cysts is determined by where the abnormal bud pinches off and whether it migrates. If pinching-off occurs near the primitive trachea, a paratracheal cyst results. Pinching-off at the tracheal bifurcation results in a carinal or hilar bronchogenic cyst. Should the abnormal budding stay in contact with the growing tracheobronchus, an intrapulmonic bronchogenic cyst results. If the pinched-off bud becomes caught between the closing sternal bars, a bronchogenic cyst results at the junction of the manubrium and the body of the sternum[21,22]. Pinching-off and migration craniad results in a bronchogenic cyst of the cervical region[8,22,23], whereas budding of the primitive lung bud without pinching-off could result in a subpleural bronchogenic cyst[8]. Migration of a bud into the primitive pericardium results in an intrapericardial bronchogenic cyst[24], whereas migration dorsally results in a para-oesophageal[25,26] or intraspinal[27] bronchogenic cyst.

Budding can occur at any time in prenatal development. If budding occurs late in pulmonary development, the resulting cyst is an alveolar or lung cyst. It can be differentiated from a bronchogenic cyst by the absence of cartilage and smooth muscle.

In summary, cysts of tracheal or bronchial origin result from abnormal

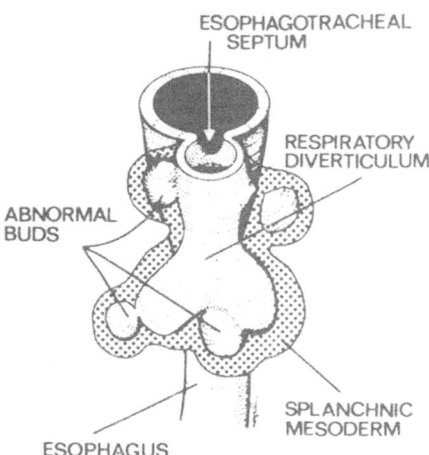

**Figure 4.2** Diagrammatic representation of abnormal budding occurring from the respiratory diverticulum. The location, pinching-off and migration of the abnormal bud determines the final location of the bronchogenic cyst

budding of the primitive tracheobronchus. Both endoderm and mesoderm must be present. Pinching-off of the abnormal bud and the extent of its migration determines the final site of the bronchogenic cyst.

## PATHOLOGY

### Location

A bronchogenic cyst can be located in any position within the chest, in the chest wall, in the neck, or under the pleura[6,8] (Figure 4.3). The most common locations are mediastinal and intrapulmonic, the latter being the most common[6–8,28]. Mediastinal bronchogenic cysts can be found in the paratracheal, hilar or subcarinal areas[6]. Not uncommonly they are located in a para-oesophageal position[25]. Intrapulmonic bronchogenic cysts are more common in the right lung, the right lower lobe being the most common site[7,8].

### Size and form

These cysts can vary from small masses of 1 cm$^2$ to large structures involving or replacing entire lobes and on occasion an entire lung[6–8]. The cysts may be single and unicystic or they may be multiple and multiloculated. The surrounding structures are not usually intrinsically abnormal, although they may appear compressed and inflamed. Communication of the cyst with surrounding structures has been described and is most common when the cyst has become infected[7,26].

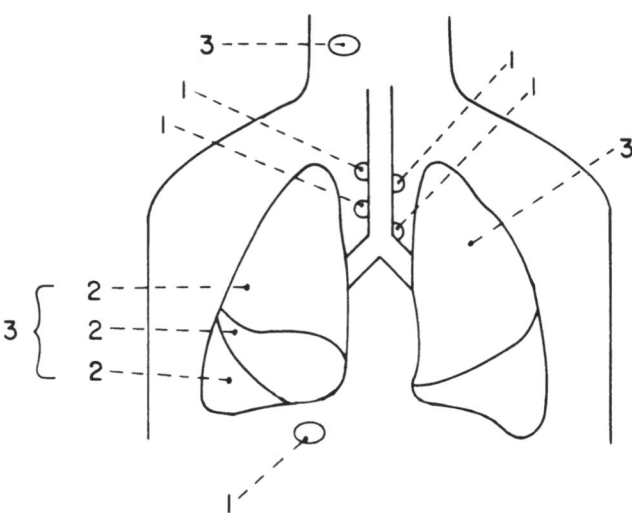

**Figure 4.3** The locations of bronchogenic cysts in a series of 20 cysts. Intrapulmonic cysts were the most common location

## Contents

Uninfected cysts located outside the pulmonary parenchyma are filled with a clear, yellowish, viscid material. If the cyst has become infected, the contents may be turbid and foul-smelling.

Intrapulmonic bronchogenic cysts are usually filled with air and fluid. Not infrequently, they will contain purulent material and the surrounding lung parenchyma will be atelectatic. Bronchiectasis may occur in the bronchioles distal to these intrapulmonic cysts because of chronic obstruction of the distal bronchi.

## Histology

Histologically, a bronchogenic cyst is lined with respiratory epithelium (ciliated columnar), and its walls contain smooth muscle, cartilage, mucous glands and elastic tissue[6-8,26]. If it has been infected, the lining membrane may have been destroyed. If the lining has been destroyed but all the other characteristic findings are present, a presumptive diagnosis of bronchogenic cyst is reasonable[26].

## NATURAL HISTORY

The natural history of a bronchogenic cyst depends upon its location, the age of the patient, the presence of infection, and expansion of the cyst.

## Location

One of the key determinants of symptoms in a patient with a bronchogenic cyst is its anatomical location. Cysts located near compressible structures frequently cause symptoms, whereas those located in areas where compression of surrounding structures does not occur may be asymptomatic.

### Mediastinal

Within the mediastinum bronchogenic cysts are found in the paratracheal, hilar or subcarinal, and para-oesophageal locations[6-8,12,14,16-18,20,25,26,29].

*Paratracheal cysts* frequently cause no symptoms because expansion of such a cyst is into the mediastinum, not toward the trachea. When expansion towards the trachea occurs, the trachea will be deviated away from the cyst[5].

*Hilar or subcarinal cysts* are most often symptomatic due to compression of the airway[5,7,12,14-16,18-20]. This results in respiratory distress in the newborn infant, frequently with hyperinflation of either lung. The hyperinflation results from a 'ball-valve' effect of the cyst, which allows free air entry but impedes the outflow of air (Figure 4.4).

*Para-oesophageal cysts*, located near or in the oesophageal wall, cause dysphagia or feeding difficulties by pressure on the oesophagus[25,26] (Figure 4.5).

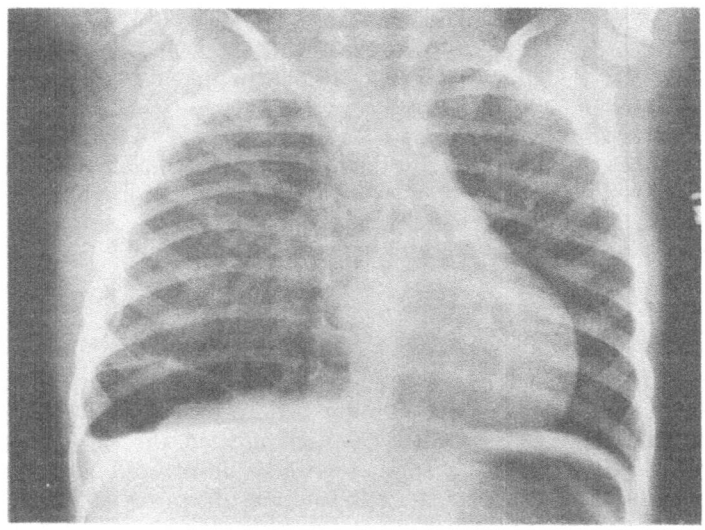

**Figure 4.4** Overinflation of the right lung of an infant due to a subcarinal bronchogenic cyst

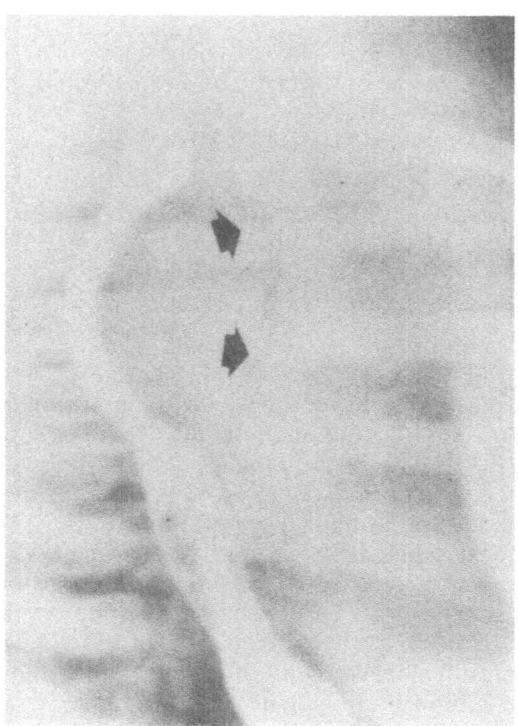

**Figure 4.5** Displacement of both the oesophagus and trachea from a para-oesophageal bronchogenic cyst. This patient's major symptom was dysphagia

## Intrapulmonic

Intrapulmonic bronchogenic cysts are the most common and are most likely to be symptomatic[7,8,17]. These cysts tend to be large, frequently are multiple, and may involve a lobe, several lobes or an entire lung (Figure 4.6). They may be located either in the central or peripheral lung fields, more commonly in the right lung, and most commonly in the lower lobe[8,12,15,19,20] (Figure 4.7).

Intrapulmonic bronchogenic cysts cause obstruction of the major bronchi and bronchioles resulting in atelectasis and pneumonia distal to the cyst[8,19,28] (Figure 4.8). Frequently the cysts themselves become infected, producing a clinical and radiological picture indistinguishable from a lung abscess[8,30]. Should rupture of the infected cyst occur, an empyema results.

## Miscellaneous

Bronchogenic cysts occur in a variety of other locations.

Cysts occurring in the *pericardium*, most often on the right side of the heart, cause symptoms by compression of the right ventricle[24,29].

*Intraspinal* bronchogenic cyst, which is most likely to occur at the C-7, 8, T-1 levels, causes numbness, tingling and weakness of the fingers or arm[27].

Cysts of the *subpleural* areas, although most often asymptomatic, are

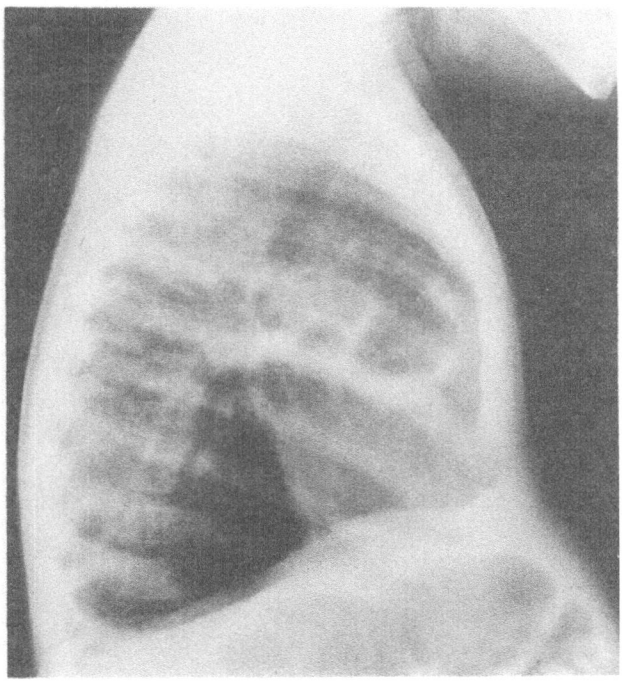

**Figure 4.6** Lateral radiograph revealing multiple bronchogenic cysts located in the right middle lobe

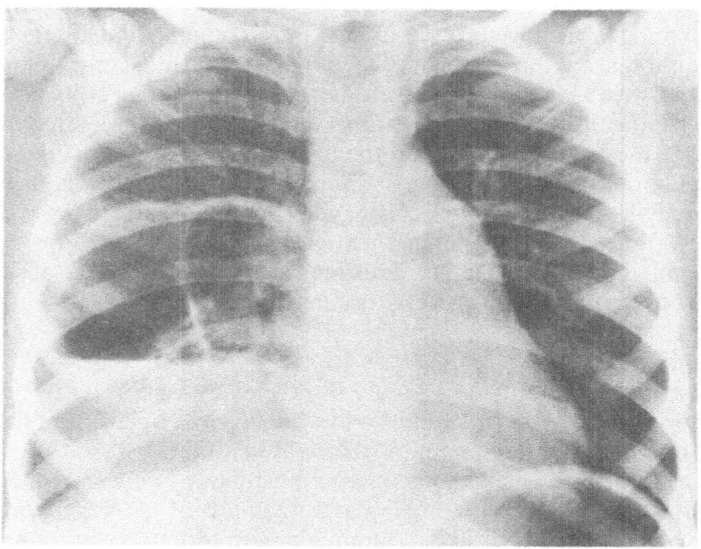

**Figure 4.7**  Two large and four smaller bronchogenic cysts containing air and fluid, located in the right middle and lower lobes

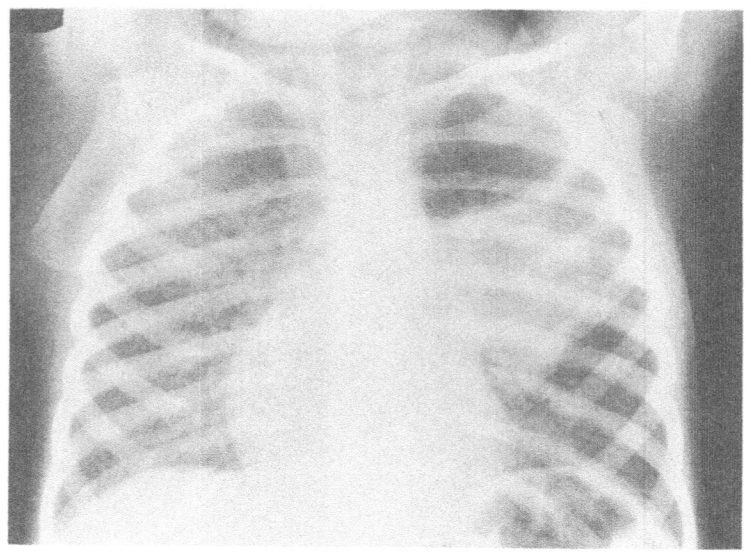

**Figure 4.8**  Pneumonia and atelectasis distal to a bronchogenic cyst located at the bifurcation of the bronchus to the left upper lobe and lingula

frequently associated with other congenital abnormalities of the primitive tracheobronchial tree such as extralobar sequestration of the lung[8,16,31]. Cysts located on the external chest wall are usually asymptomatic[21,32].

Bronchogenic cysts occurring in the *cervical region* frequently become infected but do not cause life-threatening symptoms[8,22,23]. The majority of these are clinically thought to be cysts of branchial cleft origin and are identified as bronchogenic cysts by the pathologist who finds all the component parts of a bronchus.

## Age

### Neonatal

The second determinant of symptoms is the age of the patient. In the newborn infant, bronchogenic cysts located in the hilar or subcarinal areas produce symptoms by compressing the soft, pliable airways[5,7,8,11,13-17,20]. This compression results in a respiratory distress picture frequently with hyper-expansion of the lung on the side of the lesion in the case of a hilar cyst, but of either side in the case of a subcarinal cyst (Figure 4.9). Infection of the cyst is rare in infancy.

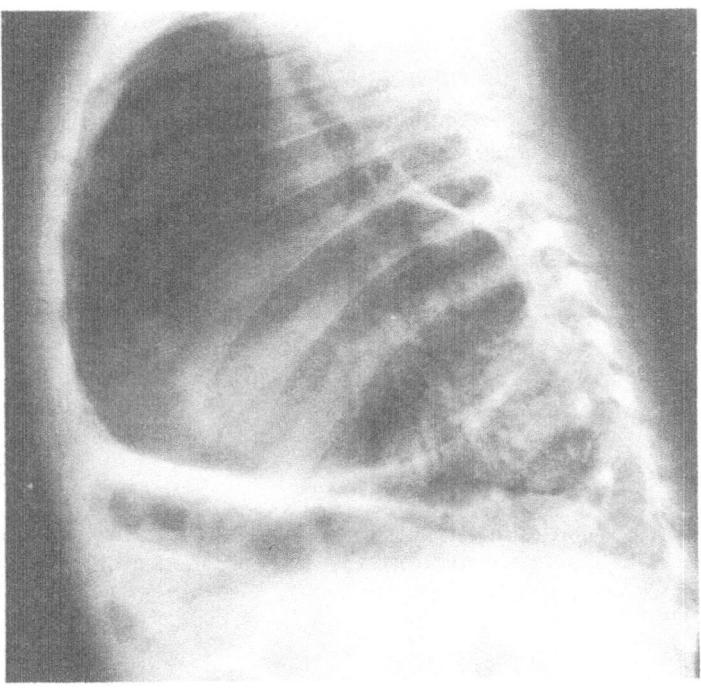

**Figure 4.9** Lateral chest radiograph with massive overinflation of the right lung due to a subcarinal bronchogenic cyst

## Childhood

Beyond the first year of life, obstruction of the airway with infection produces the majority of symptoms. Respiratory distress in the older child or adult has only rarely been reported[19]. After 10 years of age symptoms become less frequent.

## Adult

The majority of bronchogenic cysts in the adult age group are asymptomatic and when symptoms do occur they are due to infection of the cyst[19].

## Infection

The third determinant of symptoms is infection. The infection may be of the cyst itself but may also be of the distal obstructed pulmonary parenchyma. An infection in the cyst causes a clinical appearance of an intrapulmonic abscess[8,17,19,30]. Distal obstruction results in a picture of lobar pneumonia[8] (Figure 4.8). If an infected cyst ruptures, drainage may occur into the airway or into the pleural cavity (Figure 4.6). Although infection is common in intrapulmonic cysts, infection of a mediastinal bronchogenic cyst rarely occurs. Similarly, infection of a bronchogenic cyst in the neck is a frequent occurrence but infection in cysts located in other unusual locations is rare[22,23].

## Expansion

The final determinant of symptoms is expansion of the cyst. Because these cysts have mucous glands in their walls, slow expansion is to be expected. Expansion produces symptoms by compression of surrounding structures.

## CLINICAL ASPECTS

### Symptoms

The symptoms of bronchogenic cysts are diverse but patients can be divided into three groups: those with respiratory distress, those with infections and those who are asymptomatic. To a large extent, symptoms are age-related.

### Respiratory distress

This occurs primarily in infancy[5—8,12,14,17,18,20,31]. Rarely an older child or adult will manifest respiratory distress but often a chronic cough or wheezing is more common[19].

### Infection

Infection in or around a bronchogenic cyst may cause fever, bronchitis, atelectasis, recurrent or non-clearing pneumonia, recurrent lung abscess or empyema, recurrent or chronic chest pain, and haemoptysis[8,19,28]. These findings are more typical in the older child.

## *Asymptomatic*

Most bronchogenic cysts in adults are asymptomatic, being incidental findings on chest X-ray (Figure 4.10). Some authors have reported that most bronchogenic cysts are asymptomatic and are identified as incidental findings on chest radiographs. An extensive review of the literature reveals that in fact the majority of these cysts do produce symptoms[7,8,11,12,14,15,17,19,20,25,28]. In one series of 20 bronchogenic cysts in the paediatric age group, 19 were symptomatic[8]. The most common presenting symptoms were fever (6/20) and recurrent pneumonia (5/20).

The diversity of signs and symptoms makes the diagnosis of symptomatic bronchogenic cysts difficult.

## Physical examination

The examination in a patient with an intrathoracic bronchogenic cyst is rarely diagnostic. It may be normal or it may reveal various degrees of respiratory distress. The trachea may be shifted. The thorax may be dull to percussion if the cyst is large and has filled with fluid. Rales may be present on auscultation over the involved area.

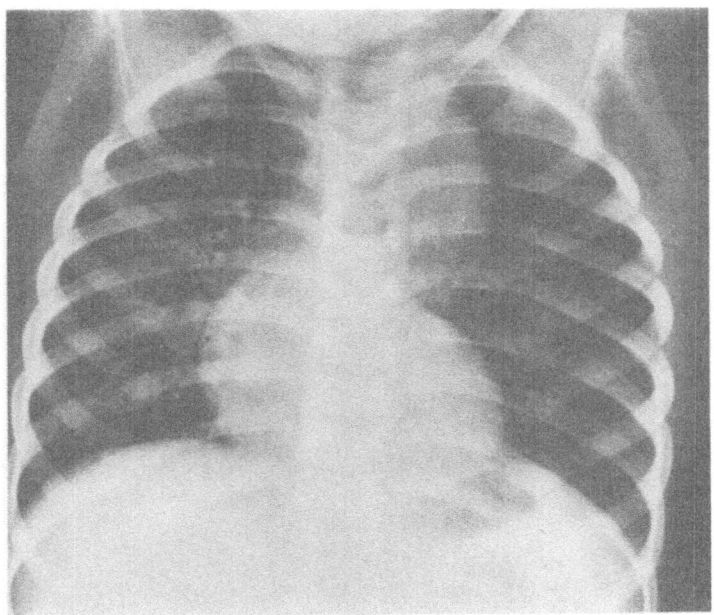

**Figure 4.10**   Chest radiograph of a patient with an asymptomatic mediastinal mass which was a bronchogenic cyst

## Diagnostic studies

A variety of diagnostic tests have been used; however, no one test is sufficient for all cases. In one reported series, although many diagnostic tests were employed, the upright chest X-ray was the most useful[8] (Table 4.1). Computed tomography (C-T) was the most accurate diagnostic test for determining the exact lesion (Figure 4.11). However, it did not change the course of therapy because a surgical lesion had been identified on chest X-ray in all patients prior to C-T scan. Bronchography is occasionally useful in patients with chronic or recurrent pneumonia due to an intrapulmonic bronchogenic cyst (Figure 4.12).

**Table 4.1  Value of various diagnostic tests used in evaluation of bronchogenic cysts[8]**

|  | Total studies | Abnormality detected % | Diagnostic of surgical lesion % | Correct diagnosis % |
|---|---|---|---|---|
| Chest X-ray | 17 | 100 | 59 | 24 |
| Barium swallow | 6 | 66 | 33 | 0 |
| Bronchoscope | 7 | 71 | 0 | 0 |
| Bronchogram | 10 | 90 | 10 | 0 |
| Angiogram | 3 | 0 | 0 | 0 |
| C-T scan | 3 | 100 | 33 | 66 |

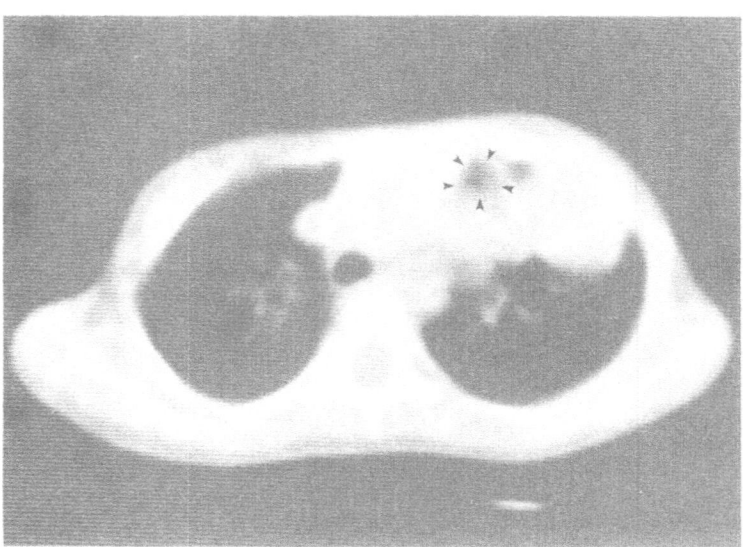

**Figure 4.11** Computer-assisted tomograph (C-T scan) demonstrating a cystic lesion (arrows) in the left upper lobe. The grey area surrounding the cyst is an inflammatory reaction and the dark area next to the cyst is an area of hyperinflation

The choice of diagnostic test should be governed by the patient's symptoms (Table 4.2). A word of caution: in a newborn with respiratory distress and hyperinflation of a lobe due to a bronchogenic cyst, bronchoscopy may be hazardous and should be performed with great caution.

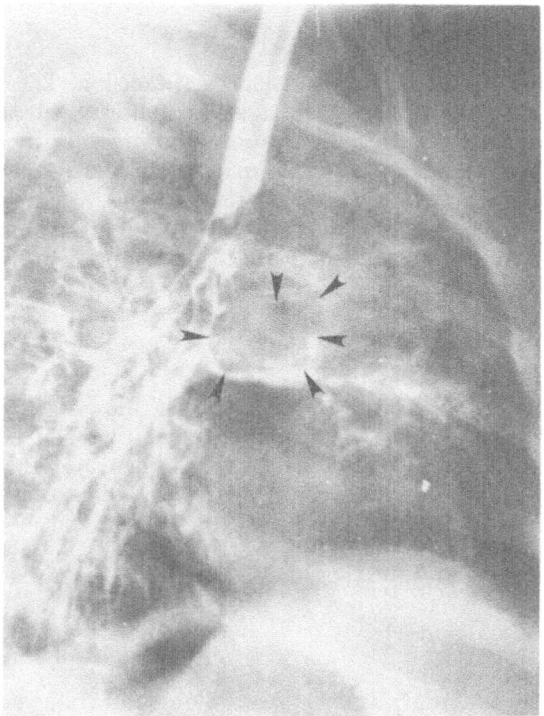

**Figure 4.12** Bronchogram demonstrating a cystic lesion in the left upper lobe which at the time of surgery was a bronchogenic cyst

**Table 4.2 Choice of diagnostic test by symptom**

| Symptom | Diagnostic examination |
|---------|------------------------|
| Dysphagia | Barium swallow |
| Cough, wheezing | Chest X-ray |
| | Bronchoscopy, Bronchogram |
| Recurrent pneumonia | Chest X-ray |
| | Chest tomography or C-T scan |
| | Bronchogram |
| Empyema, lung abscess | Chest X-ray |
| | Bronchogram |
| | C-T scan |
| Respiratory distress | Chest X-ray |
| | C-T scan |
| | ?ECHO |

M-mode ultrasonography has recently been utilized in the localization and identification of fluid-containing pulmonary cysts[33]. Further evaluation of this new method is necessary.

## Management

### Primary treatment

The treatment of bronchogenic cyst is surgical excision. The indication for surgery is the presence of the cyst, because the majority will, at some point, become symptomatic. Another indication is the presence of a mass in the mediastinum which could be a cystic lesion or a neuroblastoma or lymphoma.

The majority of bronchogenic cysts are amenable to complete excision. Rarely, partial removal with obliteration of the lining membrane may be necessary if the cyst is embedded in the wall of another organ. This has occasionally been encountered with para-oesophageal bronchogenic cysts[25,26].

### Treatment of complications

On occasion it may be necessary to treat a complication of bronchogenic cyst prior to the definitive surgery[34]. An empyema resulting from a bronchogenic cyst should be treated first with tube drainage and then, after the empyema has healed, the cyst should be removed. Similarly, an infected cyst of the neck should be drained and allowed to heal prior to removal[8,22,23]. Recurrent pneumonias resulting from partial airway obstruction should be vigorously treated prior to undertaking excision.

## Results of treatment

Using modern surgical and anaesthetic techniques, mortality is less than 1 %. Morbidity from surgical therapy is low. The morbidity and occasional mortality encountered in this congenital problem result not from the surgical treatment of the lesion but from delays in diagnosis. The most common cause of delay is difficulty in identifying a surgical lesion. In one series, failure to follow a pneumonia to complete clearing on chest X-ray resulted in long delays in diagnosis which resulted in recurring complications[8,11,17,18]. By following a pneumonia until clearing has been radiologically identified, morbidity and mortality resulting from complications of bronchogenic cysts will become extremely rare.

## References

1 Arey, L. B. (1965). *Developmental Anatomy*. 7th Edn. (Philadelphia: Saunders)
2 Gray, S. W. and Skandalakis, J. E. (1972). *Embryology for Surgeons*. (Philadelphia: Saunders)
3 Langman, J. (1975). *Medical Embryology*. 3rd Edn. (Baltimore: Williams & Wilkins)
4 Koontz, A. R. (1925). Congenital cysts of the lung. *Bull. Johns Hopkins Hosp.*, **37**, 340
5 Blackater, A. D. and Evans, D. J. (1911). A case of mediastinal cyst producing compression of the trachea, ending fatally in an infant of nine months. *Arch. Pediatr.*, **28**, 194
6 Maier, H. C. (1948). Bronchogenic cysts of the mediastinum. *Ann. Surg.*, **127**, 476

7 Rogers, L. F. and Osmer, J. C. (1964). Bronchogenic cysts: A review of 46 cases. *Am. J. Roentgenol.*, **91**, 273
8 Ramenofsky, M. L., Leape, L. L. and McCauley, R. G. K. (1979). Bronchogenic cyst. *J. Pediatr. Surg.*, **14**, 219
9 Oldham, H. N. (1971). Mediastinal tumors and cysts. *Ann. Thorac. Surg.*, **11**, 246
10 Crawford, T. J. and Cahill, J. L. (1971). The surgical treatment of pulmonary cystic disorders in infancy and childhood. *J. Pediatr. Surg.*, **6**, 251
11 Haller, J. A. Jr., Golladay, E. S., Pickard, L. R., Tepas, J. J. III, Shorter, A. M. and Shermeta, D. W. (1979). Surgical management of lung bud anomalies: Lobar emphysema, bronchogenic cyst, cystic adenomatoid malformation, and intralobar pulmonary sequestration. *Ann. Thorac. Surg.*, **28**, 33
12 Alshabkhoun, S., Starkey, G. W. B. and Asnes, R. A. (1967). Bronchogenic cysts of the mediastinum in infancy. A cause of acute respiratory distress. *Ann. Thorac. Surg.*, **4**, 532
13 Ferguson, D. F. (1970). Interesting bronchiopulmonary problems of early life. *Laryngoscope*, **80**, 1347
14 Eraklis, A. J., Griscom, N. T. and McGovern, J. B. (1969). Bronchogenic cysts of the mediastinum in infancy. *N. Engl. J. Med.*, **281**, 1150
15 Hutchin, P. (1971). Congenital cystic disease of the lung. *Rev. Surg.*, March-April, 78
16 Schmidt, F. E. and Draparras, T. (1972). Congenital cystic lesions of the bronchi and lungs. *Ann. Thorac. Surg.*, **14**, 650
17 Gourin, A., Garzon, A., Rosen, Y. and Lyons, H. A. (1976). Bronchogenic cysts: Broad spectrum of presentation. *NY State J. Med.*, May, 714
18 Whittaker, L. D. and Lynn H. B. (1973). Mediastinal tumors and cysts in the pediatric patient. *Surg. Clin. N. Am.*, **53**, 893
19 Ikard, R. W. (1972). Bronchogenic cyst causing repeated left lung atelectasis in an adult. *Ann. Thorac. Surg.*, **14**, 434
20 Gerami, S., Richardson, R., Harrington, B. and Pate, J. W. (1969). Obstructive emphysema due to mediastinal bronchogenic cysts in infancy. *J. Thorac. Cardiovasc. Surg.*, **58**, 432
21 Seybold, W. D. and Clagget, T. O. (1945). Presternal cyst. *J. Thorac. Surg.*, **14**, 217
22 Constant, E., Cavis, D. G. and Edminster, R. (1973). Bronchogenic cyst of the suprasternal area. *Plast. Reconstr. Surg.*, **52**, 88
23 Gessendorfer, H. (1973). Cervical bronchial cyst. *J. Pediatr. Surg.*, **8**, 435
24 Kwak, D. L., Storik, W. J. and Greenberg, S. D. (1971). Partial defect in the pericardium associated with a bronchogenic cyst. *Radiology*, **101**, 287
25 Suu, D. V., Carpathios, J. and Bogedain, W. (1966). Paraesophageal bronchogenic cysts: Case reports. *Am. Surg.*, **32**, 65
26 Mindelzun, R. and Long, P. (1978). Mediastinal bronchogenic cyst with esophageal communication. *Radiology*, **126**, 28
27 Yamashita, J., Maloney, A. F. J. and Harris, P. (1973). Intradural spinal bronchogenic cyst. *J. Neurosurg.*, **39**, 240
28 Gwinn, J. L., Lee, F. A., Pottenger, L. K. and Kirks, D. R. (1975). Radiological case of the month: Parenchymal bronchogenic cyst. *Am. J. Dis. Child.*, **129**, 953
29 Deenadayalu, R. P., Fisuri, D., Dewall, R. A. and Johnson, G. F. (1974). Intrapericardial teratoma and bronchogenic cyst. Review of literature and report of successful surgery in infant with intrapericardial teratoma. *J. Thorac. Cardiovasc. Surg.*, **67**, 945
30 Maier, H. C. and Haight, C. (1940). Large infected solitary pulmonary cysts simulating empyema. *J. Thorac. Surg.*, **9**, 471
31 Buntain, W. L., Isaacs, H., Jr., Payne, V. C., Jr., Lindesmith, G. G. and Rosenkrantz, J. G. (1974). Lobar emphysema, cystic adenomatoid malformations, pulmonary sequestration, and bronchogenic cyst in infancy and childhood: A clinical group. *J. Pediatr. Surg.*, **9**, 85
32 Fruga, S., Helwig, E. B. and Rosen, S. (1971). Bronchogenic cysts in the skin and subcutaneous tissue. *Am. J. Clin. Pathol.*, **56**, 230
33 Adams, F. V. and Kolodny E. (1979). M-mode ultrasonic localization and identification of fluid-containing pulmonary cysts. *Chest*, **75**, 330
34 Kirwan, W. O., Waldbaum, P. R. and McCormack, R. J. M. (1973). Cystic intrathoracic derivatives of the foregut and their complications. *Thorax*, **28**, 242

# 5
# Infantile hypertrophic pyloric stenosis: approaches to liability

J. C. BEAR

## INTRODUCTION

Among the common birth defects of man, infantile hypertrophic pyloric stenosis (IHPS) has an exceptionally straightforward and satisfactory clinical management. The epidemiology and genetics of the condition are relatively well studied, and because onset of IHPS usually occurs days or weeks after birth, several aspects of perinatal life have been investigated as possibly relevant in IHPS liability. The pathogenesis of IHPS, however, remains unknown and relatively little studied.

This review reflects the balance of investigations done and questions as yet unanswered. Inconclusive and contradictory findings are noted; areas requiring further investigation will be apparent.

As is the case for other common birth defects, much of the information available comes from a few British centres in which large numbers of index cases have been investigated in some detail. London[1,2], Belfast[3], Birmingham[4,5] and Oxford[6] studies will be mentioned repeatedly. Again typically, additional data are provided by investigations in Hawaii[7,8].

## CLINICAL ASPECTS

### Definition and pathology

Infantile hypertrophic pyloric stenosis comprises characteristic extreme hypertrophy of the pylorus, producing stenosis and vomiting sufficient to lead to detection of the hypertrophy. This qualified definition is required because the hypertrophy may never produce stenosis and symptoms[9,10], while cases with mild symptoms may remit without specific treatment[11], and thus presumably without medical ascertainment. The hypertrophy is conventionally, though loosely, termed a 'tumour'; it is not a neoplasia.

Hypertrophy involves the circular but usually not the longitudinal muscle of the pylorus. Marked dilatation of the stomach is usually found in association with IHPS; oedema of the pyloric mucosa may also occur, as may ulceration of the pyloric and occasionally the gastric mucosa[12].

## Diagnosis

In general, the pyloric hypertrophy cannot be considered congenital. In several cases, development of hypertrophy has been documented subsequent to intra-abdominal surgery for apparently unrelated disorders; in other cases hypertrophy has been surgically confirmed in the first week after birth, and was presumably congenital[10]. Most often, the diagnosis of IHPS is made after the third week but before the third month of life[3,6].

Once suspicion is aroused by persistent vomiting, which characteristically becomes projectile, IHPS is diagnosed on the observation of visible peristalsis and palpation of the tumour. When doubt remains, radiological examination will usually reveal the pyloric abnormality[12]. It has been suggested[13] that in these cases radiological examination may be avoided by examining the infant under light general anaesthesia; in a small series of cases in which IHPS was strongly suspected, tumours not otherwise palpable could consistently be felt once the infants were anaesthetized.

### Pylorospasm

Infants sometimes come under clinical observation because of severe vomiting which subsequently ceases spontaneously, or exhibit characteristic projectile vomiting without a detectable pyloric tumour. The diagnosis of pylorospasm is sometimes applied in these cases, but since undoubted pylorospasm is observable in association with other conditions[11], the additional use of the term in situations of diagnostic uncertainty seems inappropriate.

## Management

Ramstedt's operation, described in 1912, is clearly the treatment of choice for IHPS. The abdomen is opened, and the pyloric musculature split completely, down to the mucosa and along to the junction with the duodenum. The divided pyloric muscle is left unsutured, and the abdominal wound closed. If division of the musculature is incomplete, obstruction will recur, but this is rare. Vomiting should gradually subside; if severe vomiting persists for more than a week, an additional cause, or recurrent obstruction, must be suspected and dealt with if found[12]. With the application of modern fluid and electrolyte replacement, mortality from Ramstedt's operation is well below $1\%$[14–16]. Immediate resumption of normal feeding has no ill effects, and usually allows hospital discharge on the first postoperative day, thus reducing hospitalization and separation of mother and child[12,17].

Medical treatment of IHPS is possible using anticholinergic drugs, but is infrequent outside Scandinavia. The failure rate is much higher, and the fatality rate as high, as for surgical treatment, and hospitalization of the affected infant for weeks or months may be required[12].

## Complications

The only complication in the course and management of IHPS not predictable from the nature of the disorder and therapy applied is associated jaundice, which occurs in 2–3 % of cases[12,14,15]. The jaundice results from elevated serum levels of unconjugated bilirubin, and disappears after operation, thus apparently being caused by pyloric obstruction, but the mechanism of pathogenesis is unknown[3].

## Association with other disorders

The association of IHPS with other disorders has been reviewed by Dodge[10]. Association with inguinal hernia is probably a consequence of the elevation of intra-abdominal pressure with vomiting. About ten times as many affected as unaffected infants have hiatus hernia, a defect not secondary to IHPS. Associations with trisomy 18, deletion of the long arm of chromosome 21, and Turner syndrome are not surprising, considering the dysmorphological effects of chromosome abnormalities in general. The association with Turner syndrome, and the considerable excess of males affected, suggest that genes on the X chromosome reduce liability to IHPS[10].

Association with maternal myasthenia gravis could conceivably result from the effects of the cholinergic drugs used to treat that condition.

Associations with peptic and duodenal ulceration, while observed, are of uncertain significance, because the prevalence of these disorders in infancy is unknown.

Associations with rubella embryopathy and thalidomide embryopathy, with Smith–Lemli–Opitz syndrome, with de Lange syndrome, with phenylketonuria and with oesophageal atresia have not been explained.

Apart from the instances noted, the frequency of congenital malformations generally does not seem elevated among infants with IHPS[1,3,4,6,18].

## Health after infancy

The long term health of IHPS patients has not, unfortunately, been compared with that of matched controls. Follow-up of some 200 medically treated cases[19] revealed no unusual childhood or adult morbidity or mortality, and in particular no apparent excess of liability to peptic ulcer. The same appears true for surgically treated cases[11,20].

## Adult hypertrophic pyloric stenosis

The onset of hypertrophic pyloric stenosis in adulthood is difficult to diagnose without gastroscopy, but seems genuinely rare, since case reports have not increased in number with improvements in gastroscopy equipment[21]. It is commonly secondary to other gastric pathology[22]. Occasional reports[23–25] of the occurrence of adult and infantile onset hypertrophic pyloric stenosis in parents and children might well indicate some liability factors are common to both conditions, but more data are needed to clarify this link.

## ASSOCIATIONS WITH NORMAL VARIATION

### Sex

IHPS affects four or five times as many males as females, regardless of overall population incidence or race[7,26].

### Twinning

It has been suggested that twins are at increased risk of IHPS[27]. In some series excesses[6] and in other series deficiencies[2] of twin-born index cases are observed, while in yet others[1,5] about as many index cases as expected are twins. There is thus no indication that being a twin is a factor in IHPS liability.

### Physique

Carter has expressed suspicion that IHPS might be associated with muscularity[28] and large size[3]. Affected children in Belfast were not unusually tall or heavy; their muscularity was not assessed[3].

### Blood group antigens

The relative risks of IHPS for infants of blood groups O and B, as opposed to that for infants of group A, were 1.66 and 1.99 respectively in the Belfast series[29]. No associations were found with Secretor status or with Rhesus blood groups. Mothers of the Belfast index cases were even more disproportionately often of groups O and B than the patients themselves[30]. Mothers of the Oxford series of index cases, however, showed no divergence from expectation in their ABO groups, but were more often Rhesus negative than expected[6]. These slightly elevated relative risks indicate that blood group associations account for little of the differential liability of infants to IHPS[31].

## EPIDEMIOLOGY

### Incidence

*Geographical variation*
Reported incidences of IHPS are highest from Britain and northern Europe, at 2 to 4 per thousand livebirths[6,26]. Elsewhere in Europe, and among the white population of North America, the incidence appears lower, from 1 to 2 per thousand[6,26]. In the Middle East, and among Oriental and African peoples, including members of these groups residing in the continental United States and Hawaii, the incidence appears lower still, around 0.5 per thousand[6,7,26].

*Temporal variation*
  *Incidence*
There is a good evidence that the incidence of IHPS decreased by about 50% in Gothenburg between 1934 and 1959[32], and by about 30% in Belfast

between 1957 and 1969[3]. Apparent increases in IHPS incidence[4,8,33] have generally been attributed to increases in the probability of diagnosis.

*Seasonal variation*

Searches for seasonal variation in IHPS incidence have produced contradictory results. A winter peak in incidence was reported from Belfast[3], a summer peak from Oxford[6]. No evidence of seasonal variation in incidence was found in Birmingham[4], Hawaii[34], or Budapest[18].

## Association with infectious disease

There is no indication that infectious disease affecting the mother during pregnancy contributes to IHPS liability, though this has not been extensively investigated. As stated above, there are no clear indications of seasonal variation in incidence. There was no evidence, in the Oxford study, of clustering of cases in time or space[6]. When sought, no evidence was found of viral infection of the pyloric muscle of IHPS patients[35].

## Association with social class

Indications of a relationship of IHPS liability to social class are not consistent. In Belfast[3], incidence was relatively high among infants born to parents in the upper social classes, and relatively low among lower class infants. A similar, though weaker, association was found in Oxford[6]. No such association was found, though sought, in Birmingham[4] or Budapest[18]. Any tendency for upper class parents to seek more extensive medical attention for their infants could produce an apparent rising association of disease incidence with social class[10].

## ASSOCIATIONS WITH FEATURES OF PREGNANCY

### Birth order

The demonstration that a disorder is associated with order of birth is more difficult than is generally appreciated. Comparison of the birth rank distribution of affected persons with that of controls, or the general population, is only valid if sibship sizes in each group are distributed similarly[1,36,37]. Sibships of affected persons may, for several reasons, be really or apparently of different size, usually smaller, than those of controls, giving the false impression that persons of low birth rank are at increased risk. For instance:

(1) The birth of an affected child may influence parents to limit family size, to avoid having further affected children[1].

(2) Index patients are often ascertained before their sibships are complete. If so, comparisons must be drawn with controls whose sibships are similarly incomplete, and whose parents are of comparable fertility, to avoid underestimating the sibship size of the index patients[36].

95

(3) Index patients are often selected by virtue of being born in a specified time interval, usually shorter than the time it may take to complete a sibship. Ascertainment of their birth orders will thus be biased because, for sibships started during the ascertainment interval, affected members of low birth rank will be ascertained while affected members of higher ranks will not[38].

(4) The interval from the end of the specified ascertainment period to the collection of data influences the expected numbers of affected individuals ascertained, for a given sibship size, in each birth rank[37,38].

(5) A decrease in incidence or ascertainment probability during the study period will give the false impression of an association of a disorder with low birth rank[38].

IHPS is usually considered particularly common among first-born[10], following the first report on this topic by Still[39] in 1927. Still's controls were children with bronchopneumonia, which affects in particular children in large, poor families; his comparison was thus misleading[40]. There has been considerable subsequent interest in the question, with numerous investigators reporting an association of IHPS with birth rank, or its absence[6,8,10,26,37]. Studies which have taken some account of the difficulties involved[1,36—38] concur in indicating an excess of affected children among first-born, who appear perhaps twice as likely to be affected as infants in later birth ranks[1,36]. It has been suggested that females may not show this differential liability with birth order – an observation which, if confirmed, would help specify the nature of sex difference in liability[37].

## Parental age

The few investigations of this point indicate little or no association of IHPS occurrence with parental age[1,3,18,36,41].

## Birth weight

Reports from Hawaii[7] and Budapest[18] indicate that infants with IHPS average a few hundred grams heavier at birth than unaffected infants. No such difference was observed in the Oxford study[6]. A slight excess of large infants in the Belfast series[3] was thought likely to result from associations of both birth weight and IHPS incidence with social class (see above, p. 95).

## Gestation

There are several reports that IHPS is relatively infrequent in premature infants[18,42,43], but this difference has not been consistently observed[6]. Prolonged gestation did not account for the association of IHPS with elevated birth weight in the Budapest study[18]. The onset of symptoms may be earlier, the longer the gestation period[6].

## Complications during pregnancy

There are no indications that the occurrence of IHPS is associated with untoward events and complications in the course of pregnancy[3,4,6], other than maternal stress, discussed separately below.

## Emotional stress during pregnancy

Dodge[44], noting the excess of IHPS among first-born infants, and also that the first pregnancy is particularly stressful for the mother, investigated retrospectively the occurrence of stressful events in pregnancies for infants who developed IHPS. He recorded specific, unequivocally stressful events, such as the death of a first degree relative of the mother, or the unemployment of a family's principal wage earner. Among 394 mothers of IHPS patients, 25 % reported such occurrences during the third trimester of the pregnancy for the patient. By comparison, only 8 % of mothers of 62 children admitted to hospital for tonsillectomy reported such events. Stressful events in the third trimester were reported for 27 % of the pregnancies of the 101 youngest IHPS children, but for only 4 % of the pregnancies of matched controls, a statistically very significant difference. Many of the control infants for whom stress in pregnancy was reported had undiagnosed vomiting problems in early infancy.

A subsequent study[45] compared mothers of 100 IHPS patients, 100 infants who had at no time given feeding problems requiring medical advice, and 50 children with spina bifida. Mothers were given psychological inventories including the Life Events Inventory and the Eysenck Personality Inventory. Again, mothers of IHPS infants reported a striking excess of stressful events in the third trimester. Pregnancies for spina bifida children did not differ from those for normal children. The reasonable supposition that untoward events relating to any pregnancy resulting in a malformed infant would be more vividly remembered could thus be discounted. Personality inventory results showed no differences among the groups, suggesting that mothers of IHPS patients could be considered normal, non-neurotic women reacting, but not overreacting, to particularly stressful events.

## ASSOCIATIONS WITH FEATURES OF THE NEONATAL PERIOD

### Place of birth

In a Birmingham study, infants with IHPS born in hospital were observed to develop symptoms later than those born at home[46]. In the Oxford study[6] the incidence of IHPS was 2.05 per 1000 among births in large maternity hospitals but over 3 per 1000 among infants born at home or in small hospitals without obstetrical services.

### Feeding and onset of symptoms

Gerrard et al.[47], noting that the interval from birth to onset of symptoms of IHPS is about the same whether the affected infant is born prematurely or at term[42,43,47], investigated modes of infant feeding as a liability factor in IHPS. For 55 affected infants fed at 3 h intervals, average age at onset of symptoms was 21.6 d; for 97 cases fed 4 hourly, 27.1 d. This relationship could explain the later onset of symptoms in hospital- than home-delivered cases[46], since home-delivered infants were more often fed 3 hourly[47]. Possible associations of IHPS with breast feeding[3] might similarly be related to frequency of feeding.

Since timing of feeds appears to influence the onset of symptoms of IHPS, it would be worth investigating whether secular declines in IHPS incidence in some centres[3,32], and markedly lower incidences among births in large hospitals than among births in small hospitals and at home[6], are also related to changes or differences in infant feeding regimes. If appropriate feeding can reduce or prevent vomiting in some infants with pyloric hypertrophy[26,28,48], perhaps a prophylactic feeding regime could be developed to prevent IHPS in some infants at elevated genetic risk[26,49].

## Birth order and onset of symptoms

An apparent earlier onset of IHPS symptoms, the higher the birth order of the affected infant[47], may well result from increased awareness of mothers[12] and diagnosticians who have observed a previous affected child, rather than any biological cause[8]. Observations[46,49] that the excess of IHPS among first-born infants is limited to infants in whom onset of symptoms is late may simply reflect relatively late diagnosis in first-born infants superimposed on the association of IHPS with birth order.

## GENETICS

### Genetics of IHPS

Genetic investigations of IHPS are based on cases confirmed by surgical or clinical demonstration of the characteristic tumour[28], since other symptoms of the disorder are non-specific, particularly as hearsay. Observations that in pairs of monozygotic twins, one may require operation for IHPS while the other remains well, though exhibiting an elongated pylorus and incoordination of peristalsis radiologically[10] or even a palpable tumour[9], emphasize dramatically that combined genetic and environmental liability sufficient to precipitate IHPS may none the less not do so. Thus family studies provide information on the inheritance of the clinical disorder, which may differ somewhat from inheritance of pyloric hypertrophy[10].

Before Ramstedt's operation became widely known, IHPS was almost invariably fatal. Thus early family studies[1,5] could not ascertain the risk to the offspring of affected persons, though they showed the proportion of affected among sibs of index cases to be much above the general incidence among livebirths[28]. In a series of London families reported by Cockayne and Penrose[1] in 1943, about 5 % of sibs of index cases had been affected, and more parents of index cases than expected were consanguineous. This suggested recessive inheritance of IHPS, with birth order and sex influencing manifestation. The excess of males affected ruled out sex-linked dominant inheritance, and instances of male-to-male transmission ruled out sex-linked recessive inheritance. McKeown, MacMahon and Record[5], in a Birmingham series reported in 1951, observed a similar incidence among sibs, but no increase in incidence among cousins, and no increase in consanguinity among parents. Realizing that a recessive gene for IHPS liability would have to be of high frequency but low penetrance, and impressed by the association of the condition with birth order, they concluded any genetic liability to be insignificant by comparison with environmental contributions to liability.

When they found the incidence of IHPS to be of the same order among the sons and daughters of affected persons as among their brothers and sisters, Carter and Powell[49] suggested that genetic liability was more probably dominant than recessive, with penetrance greatly reduced if inheritance was monogenic. Association with birth order, and the later onset of symptoms in hospital- than home-delivered patients, were again invoked as evidence of environmental influences on the manifestation of liability.

McKeown and MacMahon[50] noted that the offspring of affected females were much more liable to IHPS than the offspring of affected males. In 33 families with an affected parent and child, they found that in 52 % (17) the mother had been affected, though only about 20 % of affected persons are female. These authors also observed that when a parent and one child were affected, the risk to subsequent children was elevated over the overall recurrence risk. These results were considered completely inconsistent with any simple genetic hypothesis, but readily explicable in terms of the effects of maternal environment.

Carter continued to monitor the offspring of IHPS patients in London, and in 1961[28] proposed a two-component genetic liability to IHPS, comprising 'first a common dominant gene, and second a sex-modified multifactorial background'. Concordance in monozygotic twins, though higher than that in dizygotic twins, was known to be only about 50 %[40], suggesting only about half of infants of predisposing genotype developed IHPS. The multifactorial component of liability was suggested to be analogous to the polygenic inheritance of height, normally distributed with male liability being considerably greater than female[28]. This model could explain why, though females were much less often affected than males, their first degree relatives were much more often affected than those of males. With polygenic inheritance of a trait, first degree relatives of affected individuals have a mean genetic liability for that trait halfway between the population mean and that of the affected individuals. As females with IHPS are relatively rare, their liability is presumably considerably more extreme than that of affected males. The liability, and proportion affected, of first degree relatives of female patients is thus expected to be greater than that of first degree relatives of male patients, as observed.

In 1969 Carter and Evans[2] reported on family data collected for nearly 1000 IHPS index patients. Their chief findings were:

(1) 20 % of the sons and 7 % of the daughters of female index patients were affected; only about 5 % of the sons and 2.5 % of the daughters of male index patients were affected.

(2) 9% of the brothers and 4% of sisters of female index patients were affected; 4% of the brothers and 2.5% of sisters of male index patients were affected.

(3) There was no indication that liability of sibs of index patients was increased when parents of index patients were consanguineous.

(4) The overall proportion affected among the nephews and nieces of index patients was five times the population incidence, with no apparent excess risk to the relatives of female index patients.

(5) The overall proportion affected among the first cousins of index patients was 1.8 times the population incidence with, again, no apparent excess risk to relatives of female index patients.

(6) When index patients had already one affected child, the recurrence risk among subsequent children was much elevated; also, the recurrence risk was much elevated when parents were unaffected but a sib of the index patient was also affected.

On the basis of these observations, particularly the low risks to second and third degree relatives[51], which ruled out any substantial contribution to liability from a dominant gene of major effect, sex-modified polygenic inheritance could be proposed as completely explaining genetic liability to IHPS. Using Falconer's[52] method, estimates were obtained of the upper limit of the heritability of IHPS, that is, the proportion of liability variation attributable to additive genetic effects. These were 76% for first degree, 27% for second degree and 50% for third degree relatives, the low estimate for second degree relatives being attributable to under ascertainment of affected aunts and uncles resulting from the strict diagnostic criteria used. The overall downward trend of heritability estimates from progressively more distant relatives was opposite to expectation, were a single dominant gene of major effect to be influencing liability. Heritability estimates were noted to be higher for relatives, particularly offspring, of female index patients[2].

Data from Belfast subsequently confirmed the London estimates of recurrence risks in sibs, and by implication that the risk to the offspring of affected females was higher than predicted on a simple multifactorial argument. Carter[53] therefore suggested an influence of maternal intrauterine environment on the fetus to explain this excess.

Dodge[10], citing the excess of affected among relatives of affected females found in both the London and Belfast studies, suggested this intrauterine component in IHPS liability might itself be heritable.

Development of the model of polygenic inheritance and multifactorial aetiology as a plausible description of liability to birth defects in man, and investigation of family data for IHPS, have in large degree proceeded in unison[2,28,40,51,53-56]. Meanwhile, possibilities were being investigated of explaining familial liability to common disorders by invoking a single locus of major effect but reduced penetrance, excluding other transmitted genetic or environmental contributions to liability, but allowing for a variable proportion of phenocopies resulting from environmental variation in liability. Over a wide range of plausible situations, the predictions of such a model and the multifactorial model were found to be indistinguishable[57]. At the same time, formal tests were being developed, allowing evaluation of the goodness of fit of family data to the multifactorial model when two or more liability thresholds are assumed, relating either to severity of disorder[58] or sex differences in liability[59]. Formal statistical evaluation of the fit of IHPS family data to various models of inheritance was thus of interest.

Kidd and Spence[57] compared fits obtainable of IHPS family data, pooled from several sources, to multifactorial and single major locus models. The best-fitting multifactorial model implied an offspring–parent correlation for liability of 0.44, in reasonable agreement with previous observation. This model, however, predicted fewer affected among offspring of affected females than were observed, giving an overall poor fit to the data. This result in essence lent statistical formality to previous observations of Carter[53]. A statistically better fit was obtained by postulating that 30% of male and 64% of female cases expressed a dominant liability allele with a population frequency of about 0.002, while the remainder of cases were phenocopies. This result was also considered unsatisfactory, since it could not account for the majority of cases of IHPS[57].

Using a test of goodness of fit to the multifactorial model, which took account of ascertainment probability, Gladstien et al.[60] were unable to reject that model as an explanation of the distribution of affected sibs of index cases in the Birmingham family data[5], but the significance of this result was unclear, since no specific alternative was tested.

Morton and MacLean[61] have developed a mixed model, which allows investigation of combinations of effects of a single locus of major effect, a polygenic component and an environmental component, all contributing independently to liability to a disorder. This model has the advantage of treating multifactorial and single major locus models as subhypotheses for direct comparison[8]. Applied[62] to pooled English IHPS family data[1,2,5], this model indicated that the contribution of a single locus of major effect, in addition to polygenic and environmental liability variation, was not necessary in order to explain the distribution of affected relatives. A maternal effect on liability was again indicated; heritability was estimated as 0.84 including, and 0.79 excluding, families with the mother affected.

In summary, both simple and complex analyses of family data suggest IHPS liability may be considered primarily genetic, involving multiple genetic and environmental factors, the influences of which are effectively cumulative and indistinguishable. No indication has been found of a single gene of major effect on liability, but children of affected mothers are at greater risk than would be predicted solely on this simplest argument of multifactorial liability.

## Familial unequal sex ratio

Knox[63], in the Birmingham data[5], noted that IHPS patients had more aunts than uncles. For index cases with cleft lip (with or without cleft palate), familial patterns of unequal sex ratio among unaffected relatives have been demonstrated[64–67]. Moreover, spontaneous abortion frequencies are reduced in sibships with more than one member affected with cleft lip. Together, these observations may indicate the effects of sex-specific, inherited factors selectively sparing male embryos in sibships with one affected, and female embryos in sibships with two affected members[67], or alternatively, the sex-specific loss of embryos so early that pregnancy is not recognized[65]. Given these findings for cleft lip, Knox's observations regarding the aunts and uncles of IHPS patients, and the very unequal sex ratio in IHPS, patterns of sex ratio variation among unaffected relatives of IHPS index cases were investigated[68]. Combined published data[1,2,5], revealed an excess of brothers over sisters among the unaffected members of sibships with more than one affected member. Contrary to Knox's[63] observations in Birmingham, the London patients of Carter and Evans[2] had more aunts than uncles; however, over both series combined, more aunts than uncles were ascertained through sibships with two affected members. No conclusions could be drawn from the limited data available on spontaneous abortions in the sibships of index patients[68]. Since IHPS seems rarely if ever congenital, it is perhaps not surprising that despite the unequal sex ratio among affected there are no indications of an association with prenatal loss.

## PATHOGENESIS

The pathogenesis of IHPS is unknown. Both pre- and postnatal factors must be suspected, because the tumour, though absent at birth, may develop soon after[11]. Few possibilities have been extensively investigated.

## Work hypertrophy

That the tumour of IHPS results from work hypertrophy due to repeated pylorospasm seems unlikely[11,12]. When the effects of confounding variables are controlled statistically, tumour size is not related to duration of symptoms but rather to age[47] and weight[3] of affected infants. Small tumours may cause obstruction while large ones may not[11], and the occurrence of tumours in the absence of symptoms may be recalled. Moreover, when the pyloric hypertrophy is bypassed by gastrojejunostomy, it nonetheless persists for many years[69,70].

## Failure of coordination

Failure of coordination of gastric emptying, that is, of pyloric relaxation with antral contraction, has long been mooted as underlying IHPS; it is postulated that pyloric hypertrophy is exacerbated by symptoms of incoordination[11]. There are several ways in which coordination might be compromised.

### Innervation defect
Defects or immaturity of pyloric innervation have been suggested in IHPS[71]. Interpretations of histochemistry[72] and ultrastructure[73] of pylorus tissue from affected infants are contradictory, the former being taken to support and the later to refute abnormality of ganglion cells. Whether any abnormalities

are primary or secondary to hypertrophy is arguable[12]. More extensive comparison with pylorus tissue of normal infants is needed to resolve this point. Neurological immaturity seems unlikely as a causal factor, since IHPS is less common in preterm than full term births[16].

## Enzyme imbalance

The association of IHPS with blood groups O and B in the Belfast series[29] has been tentatively explained in terms of gastric emptying. Persons of these blood groups have less intestinal acid phosphatase, and more intestinal alkaline phosphatase, than persons of blood groups A and AB. Persons of groups O and B absorb fat from the duodenum relatively slowly, which would tend to delay gastric emptying and might precipitate IHPS in susceptible infants[11].

## Hormone imbalance

Dodge[11,44] considered the hormone gastrin as a possible link between maternal stress and IHPS. Gastrin was known to stimulate acid and pepsin secretion by the stomach, and antral motility, and to cross the placenta in dogs[74]. With Karim[75–77], he convincingly demonstrated that chronic administration of pentagastrin, a gastrin analogue, to pregnant bitches induced IHPS in about 30% of the puppies subsequently born. Chronic administration of pentagastrin to puppies also induced IHPS. Pentagastrin also induced duodenal ulcers in bitches, and some of their puppies, and some treated pups of untreated mothers; among pups, ulceration was not necessarily associated with IHPS. To date, this is the only animal model of IHPS, and has stimulated a number of investigations of serum gastrin levels in affected infants, with very contradictory results. Several studies assert[78,79] and deny[80–82] elevated serum gastrin levels in fasting IHPS patients by comparison with unaffected controls. Likewise, heightened serum gastrin responses to protein meals in these infants has been asserted[79] and denied[81,82]. Serum gastrin levels of mothers of infants who developed IHPS have been found not elevated at delivery of those infants, by comparison with levels for mothers of normal controls[80]. Chronic administration of pentagastrin to pregnant rabbits did not induce IHPS in pups born subsequently[83].

Technical differences may underlie some of the disparities in the gastrin results in man, as may the considerable and overlapping ranges of serum levels observed in both normal and affected infants[79]. Differences among species in response to gastrin administered in pregnancy are hardly surprising. Gastrin sensitivity of affected infants at appropriate target cells, which can be expected to vary, has not been investigated[80].

A suggestion that elevated levels of cholecystokinin, rather than gastrin, might lead to IHPS[84] has not been borne out[85].

While the role of gastrin and other hormones in IHPS liability should be resolved, this should not be expected to provide a complete explanation of IHPS in man. The thrust of genetic results does not favour a single factor aetiology for IHPS. The overlap of gastrin measurements in normal and IHPS infants, even in studies in which the two groups can be differentiated, indicates considerable additional variation in liability. Further, the psychological investigations which brought gastrin under suspicion indicate a relationship between maternal response to stress in the third trimester and IHPS in offspring (p. 97); clinical investigations so far are tangential to this possibility. Maternal and fetal hormone levels have not been measured during pregnancies resulting in affected infants; practically, this would be very difficult.

## CONCLUSIONS

Factors demonstrably influencing IHPS liability are few. Sex and race figure prominently in liability variation. In British, and by implication European and North American populations, an additive familial, and presumably genetic, component in liability variation is seen. Offspring of affected mothers are at particular risk. How these factors increase liability is unknown, and they are in any case not alterable.

There are indications that it is feeding that precipitates pyloric hypertrophy and stenosis in susceptible infants, and that an appropriate infant feeding regime could influence or, less surely, prevent the occurrence of IHPS in some though not all of these infants.

The basis of liability to pyloric hypertrophy remains obscure. Fuller understanding of the development of hormonal control of the infant stomach seems a prerequisite to advances here.

The potential complexity of IHPS liability is highlighted by findings that, in a proportion of cases, stress affecting the mother during the third trimester of pregnancy may predispose infants subsequently born. Accepting the plausible suggestion that such an association could be hormonally mediated, it involves, at least, an environmental factor difficult to quantify, variable maternal emotional and hormonal responses to this environmental challenge, and variable sensitivity of the developing fetus to maternal hormone levels. Genetic variation might be expected in maternal hormonal response and fetal hormone sensitivity. The importance of this association to total population variation in IHPS liability is unknown, though it might be substantial, whether it contributes to multifactorial variation in liability or to departure from this model.

IHPS liability is clearly multifactorial not so much in the precise sense of mathematical modelling as in the more general and complex sense of involving multiple genetic and environmental factors[8]. The interaction of these factors appears considerably more subtle than is implied by mathematical modelling, experimentation, or clinical investigation undertaken to date.

### Acknowledgements

It is a pleasure to acknowledge the assistance of M. Fennessey and G. Burke in the preparation of this article, and helpful discussions with Drs J. A. Barrowman, A. J. Davis, R. H. Payne and K. B. Roberts.

### References

1 Cockayne, E. A. and Penrose, L. S. (1943). The genetics of congenital pyloric stenosis. *Ohio J. Sci.*, **43**, 1
2 Carter, C. O. and Evans K. A. (1969). Inheritance of congenital pyloric stenosis. *J. Med. Genet.*, **6**, 233

3 Dodge, J. A. (1975). Infantile hypertrophic pyloric stenosis in Belfast, 1957–1969. *Arch. Dis. Child.*, **50**, 171

4 MacMahon, B., Record, R. G. and McKeown, T. (1951). Congenital pyloric stenosis. An investigation of 578 cases. *Br. J. Soc. Med.*, **5**, 185

5 McKeown, T., MacMahon, B. and Record, R. G. (1951). The familial incidence of congenital pyloric stenosis. *Ann. Eugen.*, **16**, 260

6 Adelstein, P. and Fedrick, J. (1976). Pyloric stenosis in the Oxford Record Linkage Study Area. *J. Med. Genet.*, **13**, 439

7 Shim, W. K. T., Campbell, M. A. and Wright, S. W. (1970). Pyloric stenosis in the racial groups of Hawaii. *J. Pediatr.*, **76**, 89

8 Spence, M. A. and Gladstien, K. (1978). Pyloric stenosis and the simulation of Mendelism. In Morton, N. E. and Chung, C. S. (eds.) *Genetic Epidemiology*, pp. 331–351. (New York: Academic Press)

9 Lewis, F. L. K. (1944). Pyloric stenosis in identical twins. *Br. Med. J.*, **1**, 221

10 Dodge, J. A. (1973). Genetics of hypertrophic pyloric stenosis. *Clin. Gastroenterol.*, **2**, 523

11 Dodge, J. A. (1973). Infantile pyloric stenosis: inheritance, psyche and soma. *Ir. J. Med. Sci.*, **142**, 6

12 Dodge, J. A. (1975). The stomach. In Anderson, C. M. and Burke, V. (eds.) *Paediatric Gastroenterology*, pp. 81–123. (Oxford: Blackwell Scientific)

13 Freund, H., Berlatzky, Y., Katzenelson, R. and Schiller, M. (1976). Diagnosis of pyloric stenosis. *Lancet*, **1**, 473

14 Benson, C. D. and Lloyd, J. R. (1964). Infantile pyloric stenosis. *Am. J. Surg.*, **107**, 429

15 Schärli, A., Sieber, W. K. and Kiesewetter, W. B. (1969). Hypertrophic pyloric stenosis at the Children's Hospital of Pittsburgh from 1912 to 1967. *J. Pediatr. Surg.*, **4**, 108

16 Dodge, J. A. (1972). Infantile pyloric stenosis: a multifactorial condition. *Birth Defects: Orig. Art. Ser.*, **8**, 15 (New York: National Foundation)

17 Prosser, R. (1965). Infantile hypertrophic pyloric stenosis. *Pediatr. Surg.*, **58**, 881

18 Czeizel A. (1972). Birthweight distribution in congenital pyloric stenosis. *Arch. Dis. Child.*, **47**, 978

19 Berglund, G. and Rabo, E. (1973). A long-term follow-up investigation of patients with hypertrophic pyloric stenosis – with special reference to heredity and later morbidity. *Acta Paediatr. Scand.*, **62**, 130

20 Wanscher, B. and Jensen, H.–E. (1971). Late follow-up studies after operation for congenital pyloric stenosis. *Scand. J. Gastroenterol.*, **6**, 597

21 Dye, T. E., Vidals, V. G., Lockhart, C. E. and Snider, W. R. (1979). Adult hypertrophic pyloric stenosis. *Am. Surgeon*, **45**, 478

22 Walton, T. P. (1978). Pyloric stenosis. *Am. Surgeon*, **44**, 329

23 Fenwick, T. (1953). Familial hypertrophic pyloric stenosis. *Br. Med. J.*, **2**, 12

24 Woo-Ming, M. (1961). Familial relationship between adult and infantile hypertrophic pyloric stenosis. *Br. Med. J.*, **1**, 476

25 Zavala, C., Bolio, A., Montalvo, R. and Lisker, R. (1969). Hypertrophic pyloric stenosis: adult and congenital types occurring in the same family. *J. Med. Genet.*, **6**, 126

26 Leck, I. (1976). Descriptive epidemiology of common malformations (excluding central nervous system defects). *Br. Med. Bull.*, **32**, 45

27 Ford, N., Brown, A. and McCreary, J. F. (1941). Evidence of monozygosity and disturbance of growth in twins with pyloric stenosis. *Am. J. Dis. Child.*, **61**, 41

28 Carter, C. O. (1961). The inheritance of congenital pyloric stenosis. *Br. Med. Bull.*, **17**, 251

29 Dodge, J. A. (1971). Abnormal distribution of ABO blood groups in infantile pyloric stenosis. *J. Med. Genet.*, **8**, 468

30 Dodge, J. A. (1974). Maternal factor in infantile hypertrophic pyloric stenosis. *Arch. Dis. Child.*, **49**, 825

31 Edwards, J. H. (1965). The meaning of the associations between blood groups and disease. *Ann. Hum. Genet.*, **29**, 77

32 Walgren, A. (1960). Is the rate of hypertrophic pyloric stenosis declining? *Acta Paediatr.*, **49**, 530

33 McLean, M. M. (1956). The incidence of infantile pyloric stenosis in the north-east of Scotland. *Arch. Dis. Child.*, **31**, 481

34 Campbell, M. A. (1969). The question of seasonal variation of pyloric stenosis. *J. Pediatr.*, **74**, 1006

35 Herweg, J. C., Middlekamp, J. N., Thornton, H. K. and Reed, C. A. (1962). A search into the etiology of hypertrophic pyloric stenosis. *J. Pediatr.*, **61**, 309

36 McKeown, T., MacMahon, B. and Record, R. G. (1951). The incidence of congenital pyloric stenosis related to birth rank and maternal age. *Ann. Eugen.*, **16**, 249

37 Gladstien, K. and Spence, M. A. (1978). A statistical analysis of birth-order effects with application to data on pyloric stenosis. *Ann. Hum. Genet.*, **42**, 213

38 Barker, D. J. P. and Record, R. G., (1967). The relationship of the presence of disease to birth order and maternal age. *Am. J. Hum. Genet.*, **19**, 433

39 Still, G. F., (1927). Place-in-family as a factor in disease. *Lancet*, **2**, 795

40 Carter, C. O. (1965). The inheritance of common congenital malformations. In Steinberg, A. G. and Bearn, A. G. (eds.) *Progress in Medical Genetics*. Vol. IV, pp. 59–84. (New York: Grune & Stratton)

41 Osawa, M., Yamamoto, Y., Mitsuya, Y., Tsukamoto, A., Fukuyama, Y. and Tanaka, K. (1976). A clinical genetic study on congenital hypertrophic pyloric stenosis. *Jpn. J. Hum. Genet.*, **20**, 35

42 Henderson, J. L., Brown, J. J. M. and Taylor, W. C. (1952). Clinical observations on pyloric stenosis in premature infants. *Arch. Dis. Child.*, **27**, 173

43 Wilson, M. G. (1960). Pyloric stenosis in premature infants. *J. Pediatr.*, **56**, 490

44 Dodge, J. A. (1972). Psychosomatic aspects of infantile pyloric stenosis. *J. Psychosom. Res.*, **16**, 1

45 Revill, S. I. and Dodge, J. A. (1978). Psychological determinants of infantile pyloric stenosis. *Arch. Dis. Child.*, **53**, 66

46 McKeown, T., MacMahon, B. and Record, R. G. (1952). Evidence of post-natal environment influence in the aetiology of infantile pyloric stenosis. *Arch. Dis. Child.*, **27**, 386

47 Gerrard, J. W., Waterhouse, J. A. H. and Maurice, D. G. (1955). Infantile pyloric stenosis. *Arch. Dis. Child.*, **30**, 493

48 Thompson, W. A. and Gaisford, W. F. (1935). Congenital pyloric stenosis (with observations based on 209 consecutive cases). *Br. Med. J.*, **2**, 1037

49 Carter, C. O. and Powell, B. W. (1954). Two-generation pyloric stenosis. *Lancet*, **1**, 746

50 McKeown, T. and MacMahon, B. (1955). Infantile hypertrophic pyloric stenosis in parent and child. *Arch. Dis. Child.*, **30**, 497

51 Carter, C. O. (1969). Genetics of common disorders. *Br. Med. Bull.*, **25**, 52

52 Falconer, D. S. (1965). The inheritance of liability to certain diseases, estimated from the incidence among relatives. *Ann. Hum. Genet.*, **29**, 51

53 Carter, C. O. (1972). Genetics of infantile pyloric stenosis. *Birth Defects: Orig. Art. Ser.*, **8**, 12 (New York: National Foundation)

54 Carter, C. O. (1963). The genetics of common malformations. In *Second International Conference on Congenital Malformations*, pp. 306–313. (New York: The International Medical Congress Ltd)

55 Carter, C. O. (1970). Multifactorial inheritance revisited. In Fraser, F. C. McKusick, V. A. and Robinson, R. (eds.) *Congenital Malformations*, pp. 227–232. (Amsterdam: Excerpta Medica)

56 Carter, C. O. (1976). Genetics of common single malformations. *Br. Med. Bull.*, **32**, 21

57 Kidd, K. K. and Spence, M. A. (1976). Genetic analyses of pyloric stenosis suggesting a specific maternal effect. *J. Med. Genet.*, **13**, 290

58 Reich, T., James, J. W. and Morris, C. A. (1972). The use of multiple thresholds in determining the mode of transmission of semi-continuous traits. *Ann. Hum. Genet.*, **36**, 163

59 Kidd, K. K., Reich, T. and Kessler, S. (1973). A genetic analysis of stuttering suggesting a single major locus. *Genetics*, **74**, S137

60 Gladstien, K., Lange, K. and Spence, M. A. (1978). A goodness-of-fit test for the polygenic threshold model: application to pyloric stenosis. *Am. J. Med. Genet.*, **2**, 7

61 Morton, N. E. and MacLean, C. J. (1974). Analysis of family resemblance. III. Complex segregation of quantitative traits. *Am. J. Hum. Genet.*, **26**, 489

62 Lalouel, J. M., Morton, N. E., MacLean, C. J. and Jackson, J. (1977). Recurrence risks in complex inheritance with special regard to pyloric stenosis. *J. Med. Genet.*, **14**, 408

63 Knox, G. (1958). On the nature of the determinants of congenital pyloric stenosis. *Br. J. Prev. Soc. Med.*, **12**, 188

64 Knox, G. (1963). The family characteristics of children with clefts of lip and palate. *Acta Genet., Basel*, **13**, 299

65 Niswander, J. D., Chung, C. S., MacLean, C. J. and Dronamaraju, K. (1972). Sex ratio and cleft lip with or without cleft palate. *Lancet*, **2**, 858

66 Bear, J. C. (1973). The association of fetal wastage with facial cleft conditions. *Cleft Palate J.*, **10**, 346

67 Bear, J. C. (1978). Spontaneous abortion, sex ratio and facial cleft malformations. *Clin. Genet.*, **13**, 1

68 Bear, J. C. (1978). The association of sex ratio anomalies with pyloric stenosis. *Teratology*, **17**, 19

69 Armitage, G. and Rhind, J. A. (1951). The fate of the tumour in infantile hypertrophic pyloric stenosis. *Br. J. Surg.*, **39**, 39

70 Dickinson, S. J. and Brant, E. E. (1967). Congenital pyloric stenosis: roentgen findings 52 years after gastroenterostomy. *Surgery*, **62**, 1092

71 Friesen, S. R., Boley, J. O. and Miller, D. R. (1956). The myenteric plexus of the pylorus: its early normal development and its changes in hypertrophic pyloric stenosis. *Surgery*, **39**, 21

72 Friesen, S. R. and Pearse, A. G. E. (1963). Pathogenesis of congenital pyloric stenosis: histochemical analyses of pyloric ganglion cells. *Surgery*, **53**, 604

73 Jona, J. Z. (1978). Electron microscopic observations in infantile hypertrophic pyloric stenosis (IHPS). *J. Pediatr. Surg.*, **13**, 17

74 Bruckner, W. L., Snow, H. D. and Fonkalsrud, E. W. (1970). Gastric secretion in the canine fetus following maternal stimulation: experimental studies on placental transfer of insulin, histamine, and gastrin. *Surgery*, **67**, 360

75 Dodge, J. A. (1970). Production of duodenal ulcers and hypertrophic pyloric stenosis by administration of pentagastrin to pregnant and newborn dogs. *Nature (Lond.)*, **225**, 284

76 Karim, A. A., Morrison, J. E. and Parks, T. G. (1974). The role of pentagastrin in the production of canine hypertrophic pyloric stenosis and pyloroduodenal ulceration. *Br. J. Surg.*, **61**, 327

77 Dodge, J. A. and Karim, A. A. (1976). Induction of pyloric hypertrophy by pentagastrin. *Gut*, **17**, 280

78 Spitz, L. and Zail, S. S. (1976). Serum gastrin levels in congenital hypertrophic pyloric stenosis. *J. Pediatr. Surg.*, **11**, 33

79 Bleicher, M. A., Shandling, B., Zingg, W., Karl, W. A. and Track, N. S. (1978). Increased serum immunoreactive gastrin levels in idiopathic hypertrophic pyloric stenosis. *Gut*, **19**, 794

80 Werlin, S. L., Grand, R. J. and Drum, D. E. (1978). Congenital hypertrophic pyloric stenosis: the role of gastrin reevaluated. *Pediatrics*, **61**, 883

81 Moazim, F., Rodgers, B. M., Talbert, J. L. and McGuigan, J. E. (1978). Fasting and postprandial serum gastrin levels in infants with congenital hypertrophic pyloric stenosis. *Ann. Surg.*, **188**, 623

82 Hambourg, M. A., Mignon, M., Ricour, C., Accary, J. and Pellerin, D. (1979). Serum gastrin levels in hypertrophic pyloric stenosis of infancy. *Arch. Dis. Child.*, **54**, 208

83 Janik, J. A., Akbar, A. M., Burrington, J. D. and Burke, G. (1978). The role of gastrin in congenital hypertrophic pyloric stenosis. *J. Pediatr. Surg.*, **13**, 151

84 Rogers, I. M., Drainer, I. K., Moore, M. R. and Buchanan, K. D. (1975). Plasma gastrin in congenital hypertrophic pyloric stenosis. A hypothesis disproved? *Arch. Dis. Child.*, **50**, 467

85 Rogers, I. M., Drainer, I. K., Dougal, A. J., Black, J. and Logan, R. (1979). Serum cholecystokinin, basal acid secretion, and infantile pyloric stenosis. *Arch. Dis. Child.*, **54**, 773

# 6
# Congenital jejuno-ileal atresia and stenosis

## S. CYWES, M. R. Q. DAVIES AND H. RODE

In 1911 P. Fockens[1] of Rotterdam reported the first successfully treated case of small bowel atresia. However, up to 1952 the mortality of atresia of the small bowel remained prohibitive even at the best paediatric surgical centres in the world, viz. 84 % at the Children's Medical Center in Boston[2] and 88 % at the Hospital for Sick Children, Great Ormond Street[3] in London. Indeed, in a comprehensive review of the world's literature up to 1950, Evans[4] could find reports of only 39 cases of jejuno-ileal atresia that had been successfully treated.

Shortly after World War II Jannie Louw's first born son died of intestinal atresia. This experience and the universally poor results with treatment for this condition stimulated him to investigate this problem. In 1952[3] he published the results of an investigation of 79 patients treated at Great Ormond Street and postulated that jejuno-ileal atresia was probably due to a vascular accident rather than the result of inadequate recanalization of the bowel as suggested by Tandler[5] in 1900. At his instigation, Chris Barnard perfected the experimental model in pregnant mongrel bitches[6]. This not only confirmed the hypothesis, but also paved the way for further fetal experiments and resulted in improved survival by the clinical application of these findings. Many other workers have since repeated these experiments not only in dogs but also in rabbits[7,8] and sheep[9] and in chicken embryos[10].

During the past 25 years there has been a steady improvement in the results of treatment of atresias and stenosis of the bowel since the change in technique. This has been due to earlier diagnosis, better supportive care and refinements in technique[11]. In fact, before 1952 the mortality rate in Cape Town was 90 %. Between 1952 and 1955, 28 % of the babies could be saved. At that stage most were treated by primary anastomosis without resection. With liberal resection of the blind ends and end-to-end anastomosis, the survival rate increased to 78 % during the period 1955–1958[12]. With further improvement in care, the overall survival rate for the period 1959–1978 increased to 88 % (Figure 6.1).

With this improved initial survival rate attention has been focussed on the

109

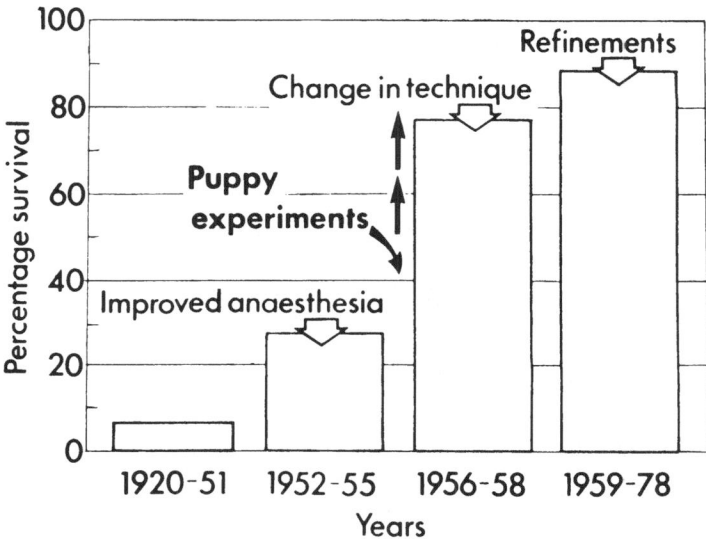

**Figure 6.1** Operative results in congenital atresia and stenosis of the small bowel at the University of Cape Town teaching hospitals

long term prognosis and results. Thus it is opportune to review the current state of thought on intestinal atresia and stenosis.

## MATERIAL

During the 20-year period 1959–1978, 84 patients with jejuno-ileal atresias and stenoses were admitted to the paediatric surgical service at the Red Cross War Memorial Children's Hospital in Cape Town.

In 1967 Louw[13] reported on the results of treatment of 33 patients with jejuno-ileal occlusions by resection of the affected segment of bowel and end-to-end anastomosis. Seven had stenosis and 26 atresias. Thirty-one of these patients survived, giving an overall survival rate of 94%.

Intestinal atresias were classified into the three types as proposed by Bland-Sutton[14] in 1889: *type I* – a simple membranous occlusion; *type II* – blind-ending bowel segments joined by a band; *type III* – disconnected bowel, ending blindly with a gap.

This simple morphological classification has stood the test of time, but since early mortality of patients born with this congenital intestinal defect has declined, the long term prognosis, which is related not only to the success of the correct surgical procedure but also to the extent of intrauterine damage to the small bowel, has become increasingly more relevant. For this reason a revised classification has been adopted as follows[15,16]. *Type I* – is the membranous (septal) occlusion, with the bowel wall continuity intact and a small intestine of normal length. *Type II* – is the cord form, in which a solid cord replaces the normal bowel anatomy and in which an associated mesenteric defect is seldom present (in rare instances the length of the small intestine may be subnormal (Figure 6.2A) ). *Type IIIa* – is the gap form, where loss of tissue of both bowel and mesentery has occurred, the unaffected intestine has an anatomically normal blood supply, and the length of the small intestine is subnormal (Figure 6.2B). *Type IIIb* – is 'apple peel' or 'Christmas tree' atresia[17,18], a gap form of jejunal atresia with an associated gross mesenteric defect, the consequence of an extensive infarction of the midgut secondary to a proximal superior mesenteric artery occlusion. The distal ileum remains viable and receives its blood supply via an abnormal arterial collateral source from the main arterial supply to the right colon. A significant loss of intestinal length always accompanies this type of atresia (Figure 6.2C). *Type IV* – the final category, comprises multiple intestinal atresias giving the appearance of a string of sausages.

Of the 84 patients currently reviewed, stenosis was the only anatomical anomaly in 17, involving the jejunum in 11 and the ileum in 6 (Table 6.1). All in this group have survived. Associated atresias were found in 3 patients and multiple areas of stenosis involving the ileum in 1 patient. There was no aetiological explanation in 13 of the patients. In 2 patients, however, a mid-gut volvulus was present and in another patient an anomaly in intestinal rotation and fixation was found. A jejunal stenosis occurred in 1 patient with mucoviscidosis.

Within the total group of jejuno-ileal occlusions, stenosis accounted for

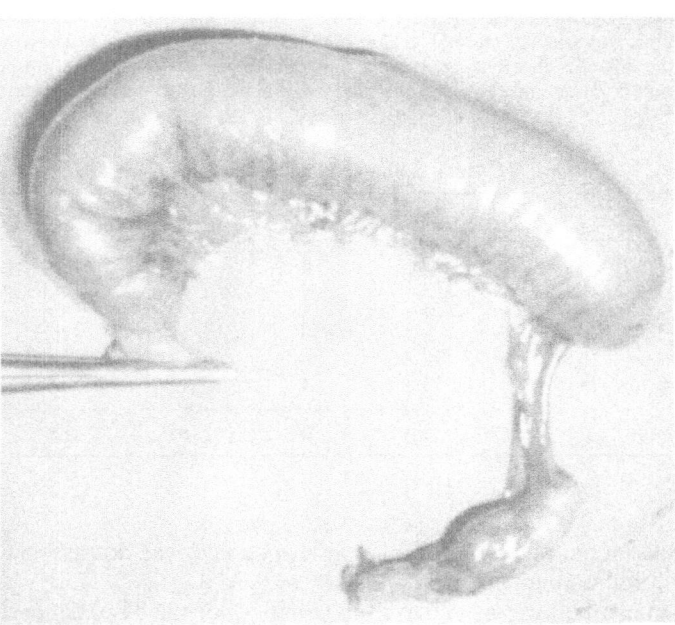

**Figure 6.2A** Atresia type II. The two blind ends are connected by a fibrous band and the intestinal mesentery is normal. This surgical specimen illustrates the extent of the proximal resection carried out on the obstructed bowel

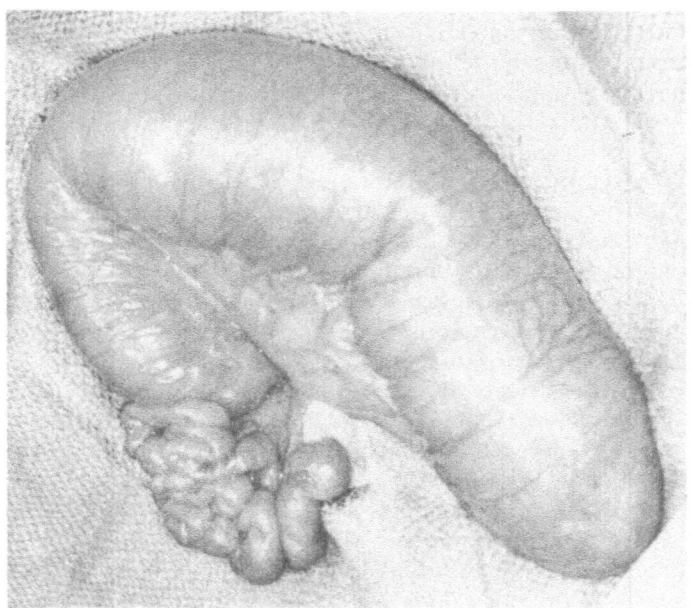

**Figure 6.2B**   Jejunal atresia type IIIa. The two blind ends of the bowel are separated by a gap and there is a defect in the mesentery. The grossly dilated obstructed bowel is seen to taper proximally into intestine of normal calibre. The collapsed and defunctioned distal bowel illustrates the difficulty in assessing the length of this bowel accurately. The presence of the mesenteric defect indicates that the total small bowel length may be subnormal

**Table 6.1   Intestinal atresia and stenosis 1959–1978**

|          |      | Jejunum | Ileum | Total | % |
|----------|------|---------|-------|-------|---|
| Stenosis |      | 11      | 6     | 17    | 20 |
| Type     | I    | 13      | 7     | 20    | 24 |
| Type     | II   | 4       | 3     | 7     | 8 |
| Type     | IIIa | 6       | 9     | 15    | 18 |
| Type     | IIIb | 8       | –     | 8     | 10 |
| Type     | IV   | 13      | 4     | 17    | 20 |
| Total    |      | 55      | 29    | 84    | 100 |

20% of all jejunal and 21% of all ileal lesions and was not associated with intrinsic colonic abnormalities.

Small-intestinal atresias occurred in 67 (80%) of the 84 patients reviewed; the number and types are detailed in Table 6.1. In 55 patients the lesions were in the jejunum and in 29 in the ileum.

During the same period we also dealt with 75 duodenal atresias in which there is a difference in the distribution of the types of lesions – blind ends being

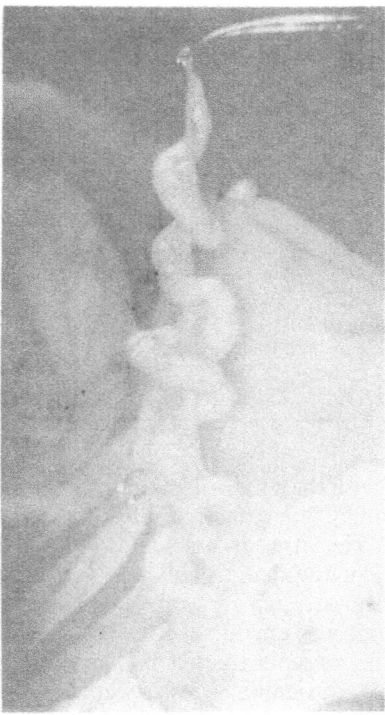

**Figure 6.2C**  Atresia type IIIb – the 'apple peel' or 'Christmas tree' form. The gross mesenteric defect always associated with this form of atresia is readily apparent

twice as common in the jejuno-ileum as in the duodenum (Table 6.2). Prematurity and severe associated anomalies are significantly more common in the duodenal occlusion group (Table 6.3).

Obvious macroscopic evidence of the prenatal vascular accident was present in 25 patients. This was the presence of a volvulus, evidence of an intussusception or snaring at the umbilical ring (Table 6.4). Significant

**Table 6.2  Types of occlusion**

|  | Duodenum (75) | Jejuno-ileum (84) |
|---|---|---|
| Stenosis | 27 } 70% | 17 } 44% |
| Type  I atresia | 25 | 20 |
| Type  II atresia | 19 } 30% | 7 } 26% |
| Type  IIIa atresia | 4 | 15 |
| Type  IIIb atresia | — | 8 } 30% |
| Type  IV atresia | — | 17 |

Table 6.3  Weight and associated anomalies

|  | Duodenum (75) | Jejuno-ileum (84) |
| --- | --- | --- |
| 2300 g | 45 (60%) | 37 (44%) |
| Severe anomalies | 38 (51%) | 15 (18%) |

Table 6.4  Evidence of a 'vascular accident' in 84 cases of jejuno-ileal occlusion

| | |
| --- | --- |
| Volvulus | 14 |
| Intussusception | 5 |
| Snaring at umbilical ring | 6 |
| Total | 25 |

anomalies associated with the jejuno-ileal occlusions were found in 15 patients and a rotational anomaly or defective intestinal fixation was noted in 29 patients (Table 6.5). The atresias associated with the omphalocoeles and gastroschisis were all in the ileum. This analysis indicates that in 57% of jejunal and 70% of ileal atresias a possible explanation for the cause of the fetal ischaemic accident was encountered.

Overall, 20% of the patients treated had multiple atresias (Type IV), a significant finding; 13 of the jejunal atresias were of this type, but only 4 of the ileal atresias were multiple. In 50% of the type IV atresias, an obvious cause or explanation of a vascular insult involving the fetal intestine was demonstrated.

During the same period, 1959–1978, an additional 4 patients with colonic atresias were seen. All were classified as type I and there were no multiple colonic occlusions. Of special interest is that in 2 patients the colonic atresia was associated with multiple small-intestinal atresias. In the remaining 2 patients the abnormalities were confined to the colon. In 1 patient an associated high anorectal abnormality was the presenting complaint.

These findings have significant implications for the surgeon performing the operative correction of any intestinal atresia, and reinforce the absolute

Table 6.5  Significant anomalies associated with 84 cases of jejuno-ileal occlusion

| | |
| --- | --- |
| Omphalocoele | 5 |
| Gastroschisis | 1 |
| Meconium peritonitis | 4 |
| Meconium ileus | 2 |
| Malrotation with volvulus | 2* |
| Duodenal atresia and Down's | 1 |
| Total | 15 (18%) |

* Twenty-nine other patients had incomplete intestinal rotation or malfixation without volvulus

**Table 6.6  Analysis of deaths**

| Age (d) | Birth weight (g) | Site | Type | Days after operation | Associated abnormalities | Cause |
|---|---|---|---|---|---|---|
| 2 | 2500 | Jejunum | I | 11 | Jaundice | Peritonitis; sepsis |
| 8 | 2700 | Jejunum | I | 19 | Nil | Intestinal obstruction; sepsis; wound disruption |
| 2 | 1800 | Jejunum | IIIb | 7 | Incomplete intestinal rotation; jaundice | Sepsis |
| 5 | 1750 | Ileum | IIIa | – | – | Aspiration during induction; no operation |
| 1 | 2500 | Jejunum | II | 2 | Meconium peritonitis | Sepsis |
| 1 | 1700 | Ileum | IIIa | – | Meconium peritonitis; moribund on admission | Perforation. No operation |
| 6 | 2040 | Jejunum | IIIb | 7 | Jaundice – exchange transfusion; pneumonia; 17 cm bowel remaining | Sepsis |
| 6 | 1800 | Jejunum | IIIb | 20 | Incomplete intestinal rotation | Sepsis |
| 5 | 1700 | Jejunum | IV | 8 | Mongolism. Duodenal atresia. Malrotation. Jaundice. 10 cm bowel remaining | Short bowel. Pneumonia |
| 3 | 2100 | Ileum | IV | 28 | Meconium peritonitis | Sepsis |

importance of assessing the patency of the lumen of the distal bowel at operation in each case.

## RESULTS

There were 10 deaths in the total group of 84, giving an overall survival rate of 88 %. The details are outlined in Table 6.6. One of these deaths was due to aspiration during induction of anaesthesia, 1 death occurred soon after admission in a moribund baby with a perforation, in whom resuscitation proved unsuccessful, and 1 was in a patient with a duodenal atresia plus multiple small-intestinal atresias and associated mongolism. This was the only patient in the series with a duodenal atresia associated with small-intestinal atresias. Five weighed 1800 g or less. In most instances the deaths were due to overwhelming infection. It is also noteworthy that the highest mortality (3 out of 8 patients) occurred in the type IIIb group (Table 6.7). In 5 of the 10, diagnosis was delayed beyond the fourth day.

Survival in relation to birth weight and associated abnormalities is depicted in Table 6.8. Fifty per cent of the jejunal atresias but only 11 % of the ileal atresias were in preterm babies. Of the associated abnormalities meconium peritonitis was the most lethal – there were 4 such infants in the series and only 1 survived.

Survival in relation to the risk group is outlined in Table 6.9.

#### Table 6.7 Mortality in various types

| Type | No. | Deaths | Mortality % |
|------|-----|--------|-------------|
| Stenosis | 17 | 0 | 0 |
| Type I | 20 | 2 | 10 |
| Type II | 7 | 1 | 14 |
| Type IIIa | 15 | 2 | 13 |
| Type IIIb | 8 | 3 | 38 |
| Type IV | 17 | 2 | 12 |

#### Table 6.8 Survival in relation to weight and associated abnormalities

| | Patients (N) | Survivors (N) | Survival rate (%) |
|------|------|------|------|
| *Weight* | | | |
| >2300/g | 47 | 44 | 94 |
| 1800 – 2200/g | 26 | 22 | 85 |
| <1800/g | 11 | 8 | 73 |
| *Associated abnormalities* | | | |
| Insignificant or | 53 | 51 | 96 |
| nil | – | 14 | 82 |
| Moderate | 17 | – | – |
| Severe | 14 | 9 | 64 |

**Table 6.9  Survival in relation to risk group**

| Risk group | Patients (N) | Survivors (N) | Survival rate (%) |
|---|---|---|---|
| A | 29 | 28 | 96.5 |
| B | 29 | 27 | 93.0 |
| C | 26 | 19 | 73.0 |
| Total | 84 | 74 | 88.0 |

## DISCUSSION

As stated before, this improvement in survival rate has been due to earlier diagnosis, better pre-, intra- and postoperative supportive care, and refinements in technique. The details of our current ideas, as well as their practical aspects, are therefore reviewed.

### Presentation and clinical features

Tremendous advances have been made in the recognition and diagnosis of intestinal obstruction in the fetus[19]. This has followed on refinements made in ultrasound evaluation of the fetus during the last 3 months of pregnancy. The obstructed and dilated fetal intestine is fluid filled and, like the fetus, surrounded by amniotic fluid, which makes it most suitable for assessment by this technique. In the past, the significance remained unrecognized in cases where an anomaly of this type was demonstrated by ultrasound. Reports are now appearing, however, in which accurate diagnosis of fetal intestinal obstruction has been made prenatally[20,21]. This often occurs accidentally as an additional finding in an investigation carried out to determine the biparietal diameter of the fetal skull, or to assess the fetus in a pregnancy complicated by polyhydramnios. Recently we have encountered two babies with intestinal atresia diagnosed prenatally on ultrasonography (Figure 6.3). In both the indication for the investigation was polyhydramnios.

The prenatal history, apart from the presence of polyhydramnios, offers little further help as a guide to the early diagnosis of jejuno-ileal atresia, as other obstetrical complications show no constant relationship to it. The family history is of some benefit[22]. Hereditary forms in which siblings have had intestinal atresias due to fetal intussusceptions have been described. An intestinal atresia may complicate mucoviscidosis, and familial forms associated with anomalies in intestinal rotation and/or fixation have also been described[23]. Finally, hereditary varieties of 'apple peel' atresia have been documented[17].

For this reason, a positive family history may precipitate early diagnostic amniocentesis. Although most atresias apparently occur during the second or third trimester, amniocentesis may detect those atresias occurring early in pregnancy, e.g. after snaring of the intestine within the extra-abdominal coelom. Raised bile salt concentration and disaccharide activity in the amniotic fluid may indicate the presence of an intestinal obstruction[24].

117

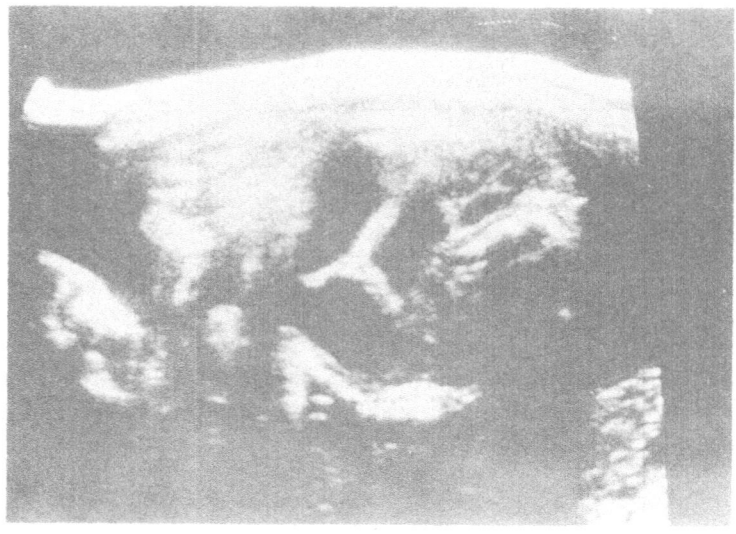

**Figure 6.3A**    Ultrasonography of fetal abdomen showing dilated loops of small bowel in the fetal abdomen

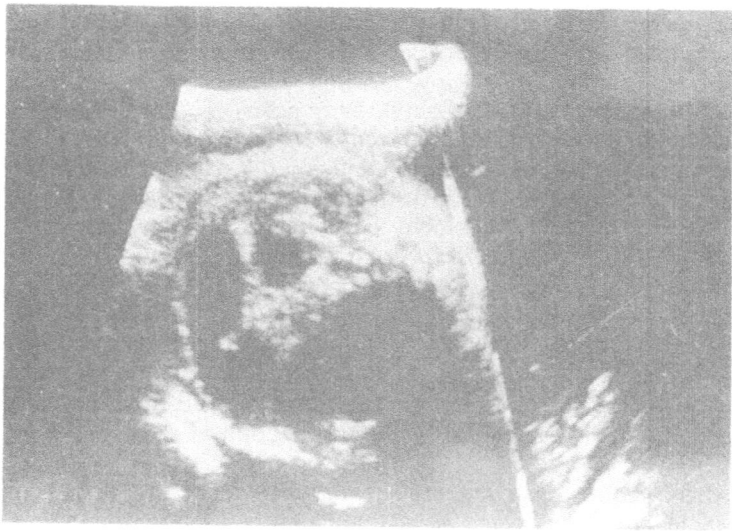

**Figure 6.3B**    Ultrasonography of fetal abdomen showing dilated stomach and dilated loops of duodenum and proximal jejunum

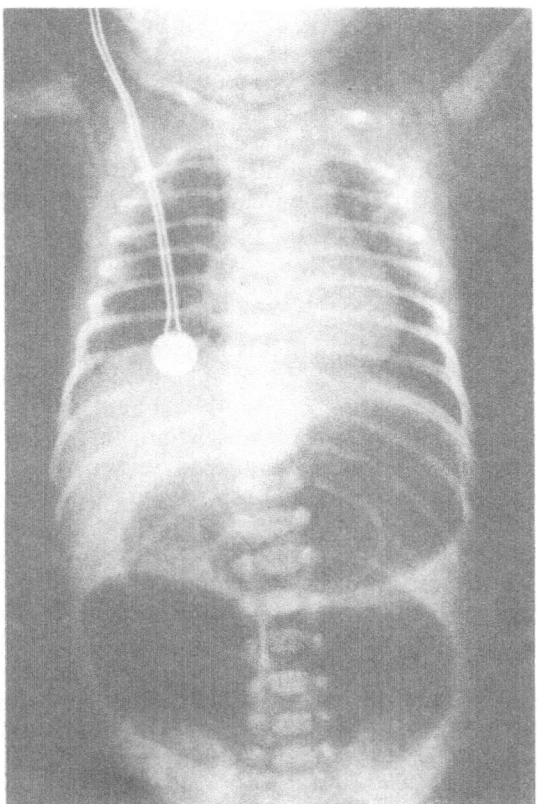

**Figure 6.3C** Radiograph of abdomen after delivery showing the dilated stomach, duodenum and proximal jejunum. The jejunal atresia was confirmed at surgery

Amniofetography may confirm this suspicion, but the effect of iodide contrast dyes on the fetal thyroid is not yet known. Between 20 % and 30 % of mothers carrying fetuses with high small-intestinal obstructions apparently develop polyhydramnios during the last trimester[25].

Forewarned is forearmed. The baby born with a congenital intestinal obstruction diagnosed *in utero* must carry a better prognosis than the unfortunate patient whose diagnosis is missed for a number of days after birth. Indeed, as suggested by Louw, intrauterine surgery may be used in the future to correct the problem before birth[26].

Nasogastric intubation of all babies born of a pregnancy complicated by polyhydramnios is routinely performed at birth. The largest, firm, red rubber catheter that can be introduced through the nostril should be used. This manoeuvre is obligatory, being first and foremost carried out to exclude oesophageal atresia. If no oesophageal anomaly is encountered, aspiration and emptying the stomach of its content could provide further clues to the

E

aetiology of the polyhydramnios. An aspirate in excess of 25 ml is regarded as indicative of a pathological abnormality. A bile-stained aspirate may suggest the presence of a high but postampullary complete small-intestinal obstruction.

Unlike babies born with high anorectal anomalies, or oesophageal or duodenal atresias[27], at least 50 % of all patients with jejuno-ileal atresia or stenosis will fall into an appropriate weight-for-age group at birth. Preterm deliveries, however, are not uncommon[16]. When an obvious congenital abnormality affects the integrity of the surrounding parietal structures enclosing the abdominal viscera, complicating intestinal atresias require exclusion. Dysmorphic anomalies of this type include exomphalos, gastroschisis and posterolateral diaphragmatic hernias.

If the diagnosis has not been made prenatally or at birth, patients are usually presented to the clinician between the 1st and 5th day of life. Classic signs of small-intestinal obstruction in the newborn are bile-stained vomiting, abdominal distension and an anomaly in colonic evacuation. This last sign is either a complete absence of meconium, or failure to evacuate meconium adequately within the first 48 h after birth.

Of these signs, vomiting of green gastric content is of great significance. In the absence of an obvious medical explanation for its cause, bilious emesis requires prompt investigation to exclude an intestinal obstruction. A high jejunal atresia is usually revealed within the 1st day of life, with vomiting the major symptom. A minor degree of upper abdominal distension and delay in passing stools may be associated findings. The degree of abdominal distension depends upon the level of the obstructing lesion; the more distal the lesion, the greater the degree of distension. On abdominal palpation, cyst-like masses representing a grossly distended proximal intestine may be felt, which may disappear after effective emptying of the stomach and proximal intestine. An infarcted mass of intestine complicating a volvulus may also be palpated, but in these instances associated clinical findings suggesting the presence of peritonitis will be found. Meconium peritonitis may be diagnosed when a meconium hydrocoele is present.

It is stated that 25 % of patients with intestinal atresias pass meconium, but in most instances the passage of the first stool is delayed beyond 24h after birth. The passage of the first stool is determined to a large degree by the volume of the meconium filling the large bowel. Intestinal content reaches the large bowel, and meconium is formed from the 4th month of intrauterine life. The meconium load contained by the large bowel depends upon when the small bowel obstructs. Experiments in animals have indicated that an infarcted fetal intestine heals, forming an atresia, within a 4 day period[8]. It follows that a colon, well filled with meconium distal to an intestinal atresia, may evacuate normally within 24 h of birth, but characteristically the first stool is usually delayed after birth. The macroscopic appearance of the meconium stool is also abnormal.

The intestinal obstruction may become complicated during the immediate postnatal period. Clinical signs of an intestinal perforation, such as a pneumoperitoneum, may be detected. This is the consequence of transmural gangrene of bowel due to tension or volvulus involving the grossly dilated proximal intestinal segment. The subtle clinical signs of a fetal (meconium) peritonitis are altered dramatically to those associated with fecal (bacterial) peritonitis.

## Diagnosis

A single abdominal radiograph with the infant erect is usually sufficient to indicate whether an intestinal obstruction is present. In congenital intestinal obstructions, after adequate emptying of the fluid-filled obstructed intestine, air is used as a contrast medium to demonstrate the level of the site of the proximal obstruction (Figure 6.4). Dilated loops of bowel with absence of gas noted in the distal intestinal tract in a normal baby 12 or more hours old suggest intestinal atresia. Occasionally, gas may be introduced into the rectum in small quantities after vigorous digital rectal examinations. Gross distension of the bowel at birth is associated with complete intestinal obstruction of some duration. In the newborn this is the sign of a fetal obstruction due to an intrinsic intestinal lesion. When an atresia is present, fluid levels are seen on the radiograph with the infant erect. Further radiographs, taken with the patient supine, prone or in an inverted lateral position, will provide further information and aid the localization and identification of the gas-filled

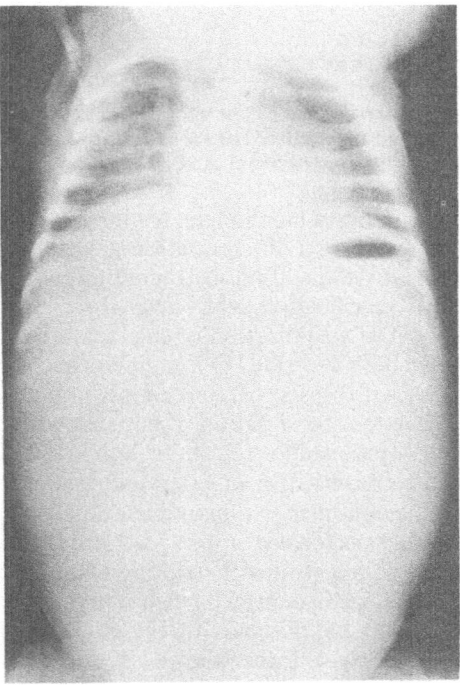

**Figure 6.4A**  An erect abdominal radiograph of a baby with an ileal atresia. The level of the obstruction cannot be determined as the intestine is filled with fluid

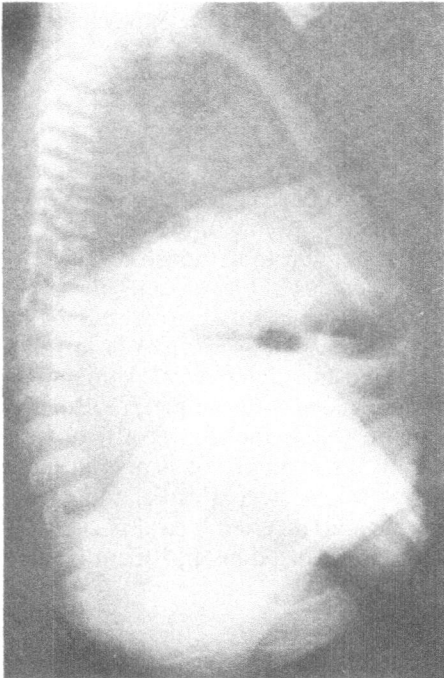

**Figure 6.4B** The erect lateral radiograph of the same patient as in Figure 5.4A, taken after the stomach had been aspirated and air introduced via the nasogastric tube. The air–fluid levels of the obstructed bowel can now be seen

intestine. Where obstruction is incomplete, for instance in the presence of an intestinal stenosis, an abnormal differentiation in size between the proximal obstructed intestine and the distal unobstructed tract will be evident. Intra-abdominal (peritoneal) calcification, which may also be intrascrotal, indicates meconium peritonitis. The visualization of calcifications within intraluminal bowel content has also been described in familial forms of intestinal atresia[28].

Radiological studies with the use of contrast medium are done for two main reasons during the diagnostic evaluation of patients with neonatal intestinal obstruction[29] – in the first instance to establish the cause and, secondly, the results may aid in the conservative or operative management. When, on a plain radiograph of the abdomen, an incomplete small-intestinal obstruction is diagnosed, a carefully performed upper gastrointestinal contrast study is indicated, which best demonstrates the site and nature of the obstructing lesion and possibly its cause. In contrast, when obstruction is complete, or the obstruction is thought to involve part of the colon or the distal ileum, a contrast enema is preferred. Either of these diagnostic studies is contraindicated before the patient is resuscitated, or when the obstruction has become complicated or meconium peritonitis is present. These studies act primarily as diagnostic aids; for example, an atresia of the terminal ileum may be confused with meconium ileus, total colonic aganglionosis, or segmental dilatation of the terminal ileum. The underlying precipitating cause of an atresia may also be demonstrated, such as an intraluminal remnant of an intussusceptum or an anomaly of rotation and/or fixation, with or without a volvulus (Figure 6.5A).

What is often not appreciated is that the barium enema study carried out on a patient with an intestinal atresia also helps the surgeon by excluding or demonstrating an unexpected associated colonic atresia (Figure 6.5B). Although this is very rare in this context, its oversight will severely jeopardize the prognosis. Neither study is without danger, since barium aspiration or perforation and peritonitis are known complications.

## Preoperative preparation

Although these patients are surgical emergency cases, time must be spent in getting the baby into optimal condition in preparation for the operation. This statement holds true even when a pneumoperitoneum is present. On the other hand, the delay must not be too long or the 'golden hour' may be passed. Furthermore, there is always the danger of ischaemia or perforation of the proximal dilated bowel.

A newborn baby precipitously evacuated from the perfectly controlled

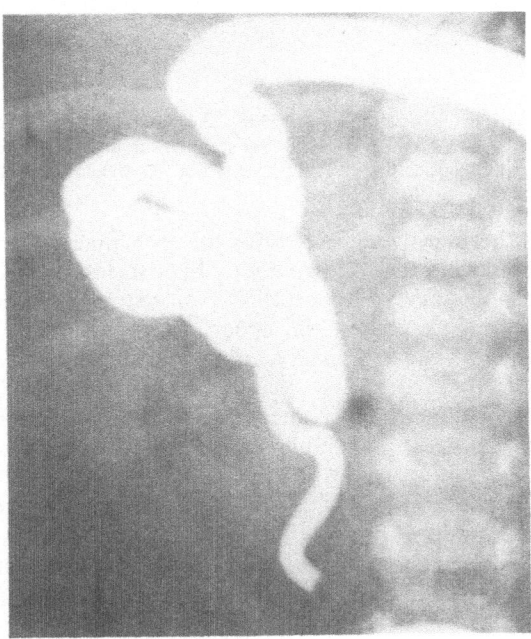

**Figure 6.5A** The diagnostic role played by a barium enema in a baby with ileal atresia. The defunctioned colon and incompletely rotated right colon are noted. Reflux of barium into the terminal ileum revealed a complete obstruction in the ileum. At laparotomy the ileal atresia was confirmed and the remnants of an intussusception were found

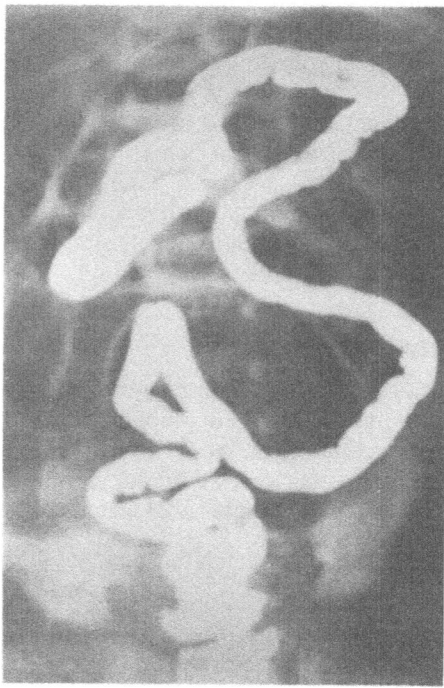

**Figure 6.5B** The barium enema study as an aid to the surgeon. The unused distal colon and a complete obstruction of the right hepatic flexure are seen. At surgery a type I atresia of the colon at that site, associated with multiple small bowel atresias, was found

environment of the uterine cavity begins life in a precarious and unstable manner. For this reason the following five 'hypostates' must be considered, corrected where necessary and controlled even before embarking on major diagnostic studies. The states, in order of importance, are hypoxia, hypovolaemia, hypothermia, hypoglycaemia and hypoprothrombinaemia.

Careful temperature control is maintained by placing the baby in a thermoneutral environment; this control must be maintained throughout the whole period of treatment. If absolute control is not strictly maintained, cold injury follows and significantly worsens the prognosis.

Blood gas levels are monitored and the fractionated inspired oxygen level is manipulated as required. It must be remembered that the presence of severe abdominal distension or even isolated, gross gastric distension itself will result in hypoventilation; thus nasogastric intubation and deflation of the upper gastrointestinal tract is mandatory. Needle aspiration of a large pneumoperitoneum in a distressed baby will achieve the same desired improvement in ventilation.

By infusing 10% dextrose in all intravenous solutions administered, hypoglycaemic episodes are prevented. In spite of this measure, regular testing with Dextrostix must be done to ensure adequate sugar levels.

Where misdiagnosis has delayed treatment, complications may have resulted. These are carefully and fully corrected before operation. Fluid and electrolyte balances are re-established, and acid-base abnormality is corrected. In the shocked baby a colloid-containing intravenous solution (blood, plasma) is administered in volume not exceeding 20 ml/kg. When calculating replacement fluid volumes, third-space losses (intestinal puddling) must be taken into account.

Just as the baby as a whole requires preparation for surgery, so the organ to be operated upon, the gastro-intestinal tract, is prepared in a similar manner. Adequate nasogastric drainage decompresses the proximal intestine and tension gangrene of the dilated proximal obstructed segment is prevented. The gastric aspirate is cultured; if Gram staining reveals multiple micro-organisms, a topical intestinal antimicrobial agent is given. A gentle rectal examination reduces anal sphincteric tone and rules out any anatomical abnormality at this site.

Finally, a therapeutic dose of vitamin $K_1$ is given. Fresh blood is cross-matched and the serum bilirubin level determined. Phototherapy is initiated when indicated and a preoperative exchange transfusion carried out if required. The operation is delayed until such time as the bilirubin level has been reduced to a satisfactory level.

If antibiotic administration has not already begun, prophylactic broad-spectrum antibiotics must be given immediately before operation. Before any antibacterial agents are given, however, all aspirated fluids, the umbilical cord, throat, nose, eyes and ears must be sampled for micro-biological examination so that the microflora are identified before antimicrobial treatment.

## Operation and intraoperative management

Although the operative management has been comprehensively described by Louw[12], the following points require reinforcement:

(1) A meticulous anaesthetic technique, with full control of the baby's body temperature, intravascular fluid and energy requirements, is essential.

(2) The umbilical cord is amputated at the navel. Where dehydration with death of the cord has already taken place, the amputation is performed through the most proximal dead part of the cord, preventing unnecessary bacterial contamination of viable tissue planes at the site. The whole of the anterior abdominal wall is prepared by cleaning the surgical field with a prewarmed antiseptic solution. The patient is isolated in gamgee tissue and covered by a large, transparent, adherent wound drape, Steridrape.

(3) An adequate supra-umbilical transverse or a right paramedian incision, according to preference, provides the desired surgical exposure.

(4) The whole of the mid-gut loop is delivered, if at all possible, in an undisturbed state. A cause for the intestinal atresia is looked for and if found corrected; for example, an incomplete intestinal rotation, with or without a volvulus, is commonly encountered. Any non-viable intestinal

remnants are resected at this stage. The lesion involving the small intestine is evaluated and the bowel distal to the atresia is assessed for the presence of any associated atresias. Distal intrinsic membranes are best detected and localized by injecting half-normal saline into the lumen of the bowel. These findings will determine the extent of the resection.

(5) Before resection, the length of the patient's small bowel has to be measured carefully. The normal length of the small intestine at birth is approximately 250 cm[30]. It is believed that the absolute minimum the neonate needs for survival is 25 cm of ileum plus the ileocaecal valve or terminal ileum and the whole of the colon, or 50 cm of jejunum with loss of the ileum, valve and part of the right colon[31,32]. No patient has survived and grown satisfactorily in our experience with a small-intestinal length of less than 50 cm. It is our opinion that the surgeon should aim to leave not less than 75 cm of small intestine after resection, when possible[33]. Based on this knowledge, certain surgical manoeuvres, in an attempt to preserve maximal intestinal length, are indicated in those patients in whom the small intestine has been grossly damaged. Usually the proximal dilated bowel is resected adequately, but under these circumstances radical resection of the proximal dilated intestine is contraindicated. As shown by Nixon[34], the grossly dilated proximal bowel is functionally ineffective and weak or non-propulsive peristaltic movements occur. For this reason this segment of bowel is resected before axial intestinal re-anastomosis, but when bowel length is critical tailoring of this segment in the form of a duodeno-jejunoplasty is advised[35]. The dilated bowel is trimmed to a lumen size of a French 22-gauge catheter. An intestinal autostapling instrument is of great assistance during this procedure. Some surgeons prefer a 'gathering' procedure of the dilated segment.

(6) An axial entero-anastomosis is the final corrective procedure once all the distal organic intestinal lesions have been surgically repaired. In 'apple peel' atresias there is an associated severe mesenteric deficiency. As the mesentery of the 'peel' segment must not be disturbed or damaged for fear of devascularizing more proximal bowel, local 'plasties' involving the intestinal wall or blind, transluminal, firm catheter rupture of simple membranes has been advocated[36].

When it has finally been established that the distal bowel lumen is patent, an end-to-back anastomosis (Denis Browne) is constructed. A single layer of Lembert full-thickness stitches of 4/0 or 5/0 arterial silk sutures is preferred, after trimming of the mucosa.

(7) Where the atresia has been repaired in the proximal jejunum, a gastrostomy is performed and the gastrostomy tube used as a conduit for the transanastomotic placement of a small Silastic feeding tube[37]. The tube is passed and placed before completing the anterior layer of the anastomosis and stabilized at the anastomotic site by a single tethering stitch, which prevents its retrograde displacement into the stomach[38].

(8) The mesenteric defects are then carefully repaired. An unfixed mid-gut loop is placed in a position of non-rotation. Where the greater omentum

has been damaged, the traumatized segment is removed to prevent the development of adhesive complications. Where soilage with infective intestinal content has occurred, the abdominal cavity is irrigated with copious amounts of normal saline or 0.01% povidone iodine. The abdominal incision is then closed in layers, using synthetic absorbable catgut or monofilament non-absorbable sutures, or with single Lembert stitches, including the peritoneum and all other musculofascial layers and excluding the subcutaneous layer and skin (Tom Jones' closure), depending on individual preference. A running, subcuticular, synthetic absorbable stitch may be used to unite the skin. Sterile plastic skin spray and a thin continuous strip of sterile Micropore complete the skin closure. Conventional wound dressings impair abdominal wall movement, and in turn diaphragmatic excursion which is one factor in the precipitation of a postsurgical hypoventilatory state.

## Postoperative management

Accurate and constant clinical monitoring by skilled nursing staff is an absolute prerequisite of successful care during all phases of postoperative management. The adjective 'constant' emphasizes the required characteristics of patient care during this period. Eight-hourly reassessment of the patient's fluid, electrolyte and acid-base states is required. Blood gases are determined regularly. Homeostasis is effectively controlled and stabilized in this manner.

Complete gastric decompression is an early essential, and is maintained at an adequate level until gastrointestinal function is re-established. Antimicrobial drugs are administered, controlling aerobic and anaerobic microorganismal growth for the first 7–8 postoperative days. An oral antifungal agent is given prophylactically in the immediate postoperative period before normal enteral alimentation.

Parenteral feeding regimens play an established role in early postoperative management. In every instance where it is certain that normal enteral alimentation will not be established satisfactorily within the first 5 postoperative days, intravenous feeding is indicated. Intravenous carbohydrate, amino acid and fat-containing solutions are introduced in a graduated manner over a period of 4 days. Peripheral venous push-in lines are used in preference to central, surgically placed venous catheters. It is aimed to have the patient on a complete parenteral feeding regimen by the 5th postoperative day. Minor complications are encountered with peripheral vein infusions when lapses in monitoring have occurred, e.g. excessive glycosuria, hyperlipaemia and tissue necrosis at the site of intravenous infusion. It must be remembered that the intravenous fat tolerance of the preterm baby is less than that of its more mature counterpart[39].

Once intestinal function has been re-established, the patient is gradually weaned from a parenteral to an enteral energy source. However, careful dietary tailoring is required, since each patient has different and individual tolerance thresholds. Based upon the known residual small-intestinal length, predictions of the degree of intestinal hypofunction can be made. Under normal circumstances there is no problem, but when gross intestinal insufficiency is expected, a hypo-osmolar elemental diet is introduced in accurately titrated volumes. Regular monitoring for clinical signs and/or biochemical evidence of intestinal overload is required. Disaccharide intolerance is well known in these patients. This and rare monosaccharide intolerance, an indication of gross brush-border malfunction of the intestine, should be detected before severe clinical signs are demonstrated by the regular biochemical assessment of stool fluid samples. A falling pH and an associated increasing level of reducing substances in this fluid denote unsatisfactory carbohydrate assimilation.

The patient's oral intake is gradually increased in volume and energy content while the intestinal tract is allowed to adapt till maximum intake tolerance is reached, which is age determined[40]. This phase of management is approached towards the end of the first postoperative month. Daily audits of energy intake versus expected requirements are carried out and serve to reveal any energy shortfall which may be present. Rapid growth in brain size occurs during the first 6 months of life, but does not occur in the nutritionally depleted patient. Only when the energy intake is closely monitored can a complication of this type be predicted and thereby prevented. Anthropometric monitoring is a useful guide and an essential feature of this phase of management. Any degree of alimentary insufficiency resulting in a prolonged period of inadequate energy intake will require a supplemental, intravenous nutritional regimen which may necessitate a prolonged period of in-hospital care.

Episodes of acute intestinal insufficiency are common in patients with subminimal intestinal function, especially in cases where a short length of small bowel has undergone a process of gross and maximal adaptive dilatation. In these instances, decompensation of an acute nature in intestinal function is usually the consequence of inadequate movement of intestinal contents. Intraluminal stasis results in abnormal growth of enteral organisms, with damage to the mucosal surface of the bowel. Bile constituents are broken down and their products have a purgative effect on the colon. Episodes of this nature may be controlled by the use of topical, enteral antimicrobial agents or cholestyramine. Indicator dyes determine intestinal transit times and are used to monitor the effect of therapy. Pharmacologically induced control of intestinal peristaltic activity has become more effective since the introduction of loperamide hydrochloride, which has proved of value in the management of these patients. The terminal ileum is an area of specific mucosal receptor sites required for the absorption of vitamin $B_{12}$ and bile salts. Colonic irritation and deficiency states complicate resection of this part of the small intestine. Thus, vitamin supplementation as well as trace elements is required.

Maximal functional adaptation should be achieved between the 6th and 12th postoperative months. At this stage patients can be classified into one of two main groups. Patients with uncorrected alimentary insufficiency are placed in the first group, i.e. the patient manifests symptoms of the short-bowel syndrome. They require prolonged intravenous nutritional supplementation and every attempt is made to stimulate and allow maximal adaptation of the intestinal mucosa to continue. In this regard, the important role played by surface (topical) feeding is emphasized[41]. In the absence of oral nutritional intake, intestinal mucosal adaptation will remain incomplete and indeed mucosal cells may atrophy[42].

The second group is divided into three subdivisions: (1) patients with adequate alimentary function for nutrition (i.e. survival) only, (2) those with adequate alimentary function for nutrition and growth and (3) those with good alimentary function for nutrition and growth, which is associated with a degree of intestinal reserve capacity.

In predicting the ultimate functional outcome, the following factors must be taken into consideration: the ileum adapts to a greater degree than the jejunum, the neonatal small bowel, unlike its adult counterpart, still has a period of maturation and growth ahead of it and the actual residual small-bowel length is difficult to determine in the newborn. The proximal obstructed bowel segment is elongated and stretched and its length may lead to overestimation of its functional capability. In contradistinction, the defunctioned, unused distal bowel is collapsed. This portion of bowel, owing to its unused state at the time of operation, should have its length multiplied by a factor of at least two during the final calculations, when considering the actual residual bowel length present.

## COMMENT

From the foregoing it will be noted that the beneficial results obtained are due to the clinical application of the experimental findings coupled with the better supportive care. The pre-, intra- and postoperative treatment have been outlined in detail as we believe that it is this constant attention to detail, based on a better understanding of the pathophysiology of this condition, which have helped considerably in obtaining the results shown in Figure 6.1.

## References

1 Fockens, P. (1911). Ein operativ geheilter Fall von Kongenitaler Dündarmatresie. *Zentralbl. Chir.*, **38**, 532
2 Ladd, W. E. and Gross, R. E. (1941). *Abdominal Surgery in Infancy and Childhood*. (Philadelphia: Saunders)
3 Louw, J. H. (1952). Congenital intestinal atresia and severe stenosis in the newborn. *S. Afr. J. Clin. Sci.*, **3**, 109
4 Evans, C. H. (1951). Atresias of the gastrointestinal tract. *Int. Abstr. Surg.*, **92**, 1
5 Tandler, J. (1900). Zur Entwicklungsgeschichte des menschlichen Duodenum frühen Embryonalstadien. *Morphol. Jahrb.*, **109**, 187
6 Louw, J. H. and Barnard, C. N. (1955). Congenital intestinal atresia. *Lancet*, **2**, 1065
7 Blanc, W. A. and Silver, L. A. (1962). Intrauterine abdominal surgery in the rabbit foetus: Production of intestinal atresia. *Am. J. Dis. Child.*, **104**, 118
8 Tsujimoto, K., Sherman, F. E. and Ravitch, M. M. (1972). Experimental intestinal atresia in the rabbit fetus. Sequential pathological studies. *Johns Hopkins Med. J.*, **131**, 287
9 Abrams, J. S. (1968). Experimental intestinal atresia. *Surgery*, **64**, 185
10 Tibboel, D., Molenaar, J. C. and Nie, C. J. (1979). New perspectives in fetal surgery: The chicken embryo. *J. Pediatr. Surg.*, **14**, 438
11 Louw, J. H. (1966). Jejunoileal atresia and stenosis. *J. Pediatr. Surg.*, **1**, 8
12 Louw, J. H. (1959). Congenital intestinal atresia and stenosis in the newborn. Observations on its pathogenesis and treatment. *Ann. R. Coll. Surg. Engl.*, **25**, 209
13 Louw, J. H. (1967). Resection and end-to-end anastomosis in the management of atresia and stenosis of the small bowel. *Surgery*, **62**, 940
14 Bland-Sutton, J. (1889). Imperforate ileum. *Am. J. Med. Sci.*, **98**, 457

15 Martin, L. W. and Zerella, J. T. (1976). Jejunoileal atresia: A proposed classification. *J. Pediatr. Surg.*, **11**, 399

16 Touloukian, R. J. (1978). Intestinal atresia. *Clin. Perinatal.*, **5**, 3

17 Blythe, H. and Dickson, J. A. S. (1969). Apple peel syndrome (congenital intestinal atresia). A family of seven index patients. *J. Med. Genet.*, **6**, 275

18 Santulli, T. V. and Blanc, W. A. (1961). Congenital atresia of the intestine: Pathogenesis and treatment. *Ann. Surg.*, **154**, 939

19 Duenhoelter, J. H., Santos-Ramos, R., Rosenfeld, C. R. *et al.* (1976). Prenatal diagnosis of gastrointestinal tract obstruction. *Obstet. Gynecol.*, **47**, 618

20 Lee, T. G. and Warren, B. H. (1977). Antenatal ultrasonic demonstration of fetal bowel. *Radiology.*, **124**, 471

21 Fogel, S. R., Katragadda, C. S. and Costin, B. S. (1980). New ultrasonographic finding in a case of fetal jejunal atresia. *Texas Med.*, **76**, 44

22 Mishalany, H. G. and Najjar, F. B. (1968). Familial jejunal atresia: Three cases in one family. *J. Pediatr.*, **73**, 753

23 Garance, P. H., Blanchard, H., Collin, P. P. *et al.* (1973). Multiple atresias and a new syndrome of hereditary multiple atresias involving the gastrointestinal tract from stomach to rectum. *J. Pediatr. Surg.*, **5**, 633

24 Délèze, G., Sidiropoulos, D. and Paumgartner, G. (1977). Determination of bile acid concentration in human amniotic fluid for prenatal diagnosis of intestinal obstruction. *Pediatrics*, **59**, 647

25 De Lorimier, A. A., Fonkalsrud, E. W. and Hays, D. M. (1969). Congenital atresia and stenosis of the jejunum and ileum. *Surgery*, **65**, 819

26 Louw, J. H. (1970). The scientific method in surgery. *Trans. Coll. Med. S. Afr.*, **14**, 9

27 Cozzi, F. and Wilkinson, A. W. (1969). Intrauterine growth rate in relation to anorectal and oesophageal anomalies. *Arch. Dis. Child.*, **44**, 59

28 Martin, C. E., Leonidas, J. C. and Armoury, R. A. (1976). Multiple gastrointestinal atresias, with intraluminal calcifications and cystic dilatation of bile ducts: A newly recognized entity resembling 'a string of pearls'. *Pediatrics*, **57**, 268

29 Cremin, B. J., Cywes, S. and Louw, J. H. (1973). *Radiological Diagnosis of Digestive Tract Disorders in the Newborn*, p. 90. (London: Butterworths)

30 Reiquam, C. W., Allen, R. P. and Akers, D. R. (1965). Normal and abnormal small bowel lengths: An analysis of 389 autopsy cases in infants and children. *Am. J. Dis. Child.*, **109**, 447

31 Wilmore, D. W. (1972). Factors correlating with a successful outcome following extensive intestinal resection in newborn infants. *J. Pediatr.*, **80**, 88

32 Reid, I. S. (1975). The significance of the ileocaecal valve in massive resection of the gut in puppies. *J. Pediatr. Surg.*, **10**, 507

33 Rickham, P. P. (1967). Massive small intestinal resection in newborn infants. *Ann. R. Coll. Surg. Engl.*, **41**, 480

34 Nixon, H. H. (1955). Intestinal obstruction in the newborn. *Arch. Dis. Child.*, **30**, 13

35 Thomas, C. G. (1969). Jejunoplasty for the correction of jejunal atresia. *Surg. Gynecol. Obstet.*, **129**, 545

36 El Shafie, M. and Rickham, P. P. (1970). Multiple intestinal atresia. *J. Pediatr. Surg.*, **5**, 655

37 Wilkinson, A. W., Hughes, E. A. and Stevens, L. H. (1965). Neonatal duodenal obstruction: The influence of treatment on the metabolic effects of operation. *Br. J. Surg.*, **52**, 410

38 Coln, D. and Cywes, S. (1977). Simultaneous drainage gastrostomy and feeding jejunostomy in the newborn. *Surg. Gynecol. Obstet.*, **145**, 594

39 Forget, P. P., Fernandes, J. and Haverkampbegemann, P. (1975). Utilization of fat emulsion during total parenteral nutrition in children. *Acta Paediatr. Scand.*, **64**, 377

40 Voitk, A. J., Echave, V., Brown, R. A. *et al.* (1973). Use of elemental diet during the adaptive stage of short gut syndrome. *Gastroenterology*, **65**, 419

41 Levine, G. M., Deren, J. J. and Yezdimir, E. (1976). Small bowel resection: oral intake is the stimulus for hyperplasia. *Am. J. Dig. Dis.*, **21**, 542

42 Gas, N. and Noailliac-Depeyre, J. (1976). Studies on intestinal epithelium involution during prolonged fasting. *J. Ultrastruct. Res.*, **56**, 137

# 7
# Omphalocoele and gastroschisis

## J. A. NOORDIJK

Medical literature has often been vague in characterizing omphalocoele and gastroschisis, while the nomenclature has been inconsistent. Both words have been used indiscriminately, as well as others such as exomphalos, to describe what is now commonly understood to be omphalocoele.

Omphalocoele and gastroschisis are in fact two quite distinct and separate entities, clinically as well as pathologically, while there is probably also an embryological and genetic difference. The only common feature is their location at the site of the umbilicus.

To add to the confusion, omphalocoele and gastroschisis are often grouped together in the literature and this chapter is no exception to the rule.

It has rightly been suggested that another word should be found for gastroschisis, as the Greek word *gaster* is commonly used to denote stomach only and, whatever gastroschisis may entail, it certainly never amounts to a stomach fissure. Müntener[1] has suggested the designation 'para-umbilical abdominal defect' and the name 'para-omphalocoele' has also been proposed. We will use the words 'omphalocoele' and 'gastroschisis' because that designation has been generally adopted. Neither malformation is very common. The incidence of omphalocoele is about 1 : 3500 births and of gastroschisis 1 : 5–6000.

## OMPHALOCOELE

Omphalocoele is a midline defect in the abdominal wall, through which the intestines protrude. These are always covered by a membranous sac. This sac may have ruptured during birth, but sac remnants are *always* noted. In most omphalocoeles the sac is still intact and as a rule the umbilical cord emerges from the top of the sac (Figures 7.1, 7.2). The abdominal wall defect is usually large and the hernia contains not only bowel, but often other viscera as well, such as stomach, spleen and portions of the liver.

Embryologically it is a 'middle coelosomia'[2]. In rare cases the defect extends upwards into a sternal and diaphragmatic defect with an ectopia

131

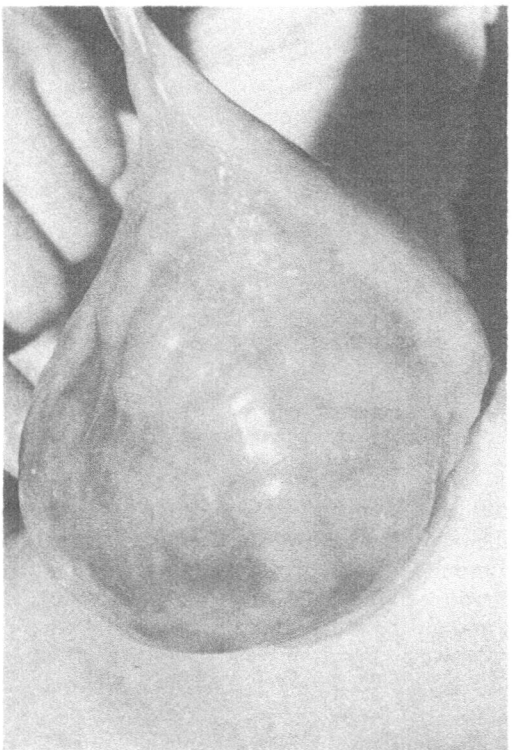

**Figure 7.1** Omphalocoele. The umbilical cord emerges from the top of the coele

cordis (upper coelosomia) or it may extend downwards and result in a partial hindgut agenesis and an open bladder (lower coelosomia). Omphalocoeles are often associated with gastrointestinal, urogenital, cardiovascular and skeletal anomalies.

## GASTROSCHISIS

In gastroschisis the defect is nearly always situated to the right of the umbilicus and the umbilical cord emerges at its normal site (Figures 7.3, 7.4). There is *never* any sign of a membranous sac. As a rule the defect in the abdominal wall is much smaller than in omphalocoeles and, other than bowel, no viscera protrude from the defect. Gastroschisis may be classified as antenatal or perinatal, depending on the aspect of the bowel. In the antenatal form the protruding bowel is thickened, oedematous and rigid and has acquired a yellow, gelatinous covering due to longstanding exposure to amniotic fluid. In the perinatal form the bowel may look almost normal. Association with other serious congenital anomalies is very rare and usually restricted to atresia and/or nonrotation of the bowel.

**Figure 7.2** This seventeenth century illustration of a case of omphalocoele appears quite elegant. (*From* Job van Meekeren (1611–1660), surgeon in Amsterdam)

## EMBRYOLOGY

It is hardly surprising that omphalocoele and gastroschisis, which exhibit so many clinical differences, should have a different developmental history.

Duhamel[2] gives a clear description of the normal embryology of the abdominal wall and on this basis he tries to explain the differences between omphalocoele and gastroschisis. In its initial developmental phase, once the development of the germinal disc is completed, the primitive embryo consists of three germ layers: the dorsal ectoblast, the ventral entoblast and, in between, the mesoblast. From the mesoblast the axial skeleton (notochord) develops into the median plane, while peripherally the embryonic and the extra-embryonic mesenchyme are formed. The notochord will eventually disappear, but it determines the differentiation of the *ectoblast* into the nervous system, of the *entoblast* into the intestinal tube and of the *mesoblast* into the skeletal, muscular and urogenital systems. From the mesenchyme adjacent to the ectoblast a dorsal somatic layer (somatopleure) is formed, while a ventral splanchnic layer (splanchnopleure) originates from the mesenchyme adjacent to the entoblast.

As the dorsal axis of the embryo grows faster than the rest, it becomes curved and elevated. This causes a doubling back of the four walls. The resulting two lateral folds will form the ventral wall together with the cephalic and caudal folds.

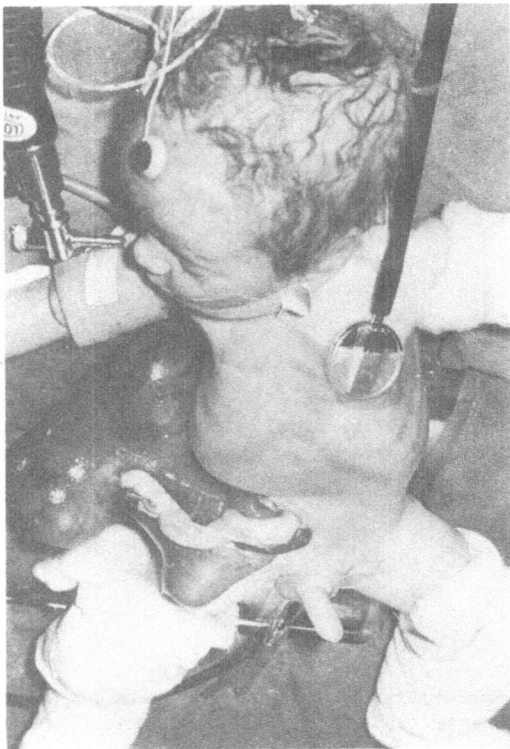

**Figure 7.3** Gastroschisis. The umbilical cord emerges at its normal place to the left of the defect

Müntener[1] raises doubts as to the existence of a cephalic fold. He is of the opinion that the abdominal wall of the supra-umbilical region is of a secondary origin.

The somatic layers of these folds form the ventral wall, whereas the gut originates from the splanchnic layers. The closure of the ventral wall around the body stalk takes place during the sixth week of pregnancy and marks the site of what will eventually become the umbilicus. The body stalk will become the umbilical cord, which initially contains a right and a left umbilical vein. At that moment morphogenesis is completed.

Inhibition of these processes by a teratogenic action, before the sixth week of gestation, will therefore produce morphological abnormalities which we define as monstrosities. Omphalocoeles belong to this category. Such an abnormal development does not prevent the secondary differentiation into various tissues during the organogenetic phase.

Should a teratogenic action occur between the sixth and the eighth week of gestation preventing the differentiation of mesenchyme into the somatopleure, then the ectoblastic layer, lacking support, will be resorbed in the same way as the oral and cloacal membranes. If this takes place in a lateral

**Figure 7.4** In this seventeenth century illustration the neonate with gastroschisis tries to push back the protruding intestines with its left hand; (*From* Job van Meekeren (1611–1660), surgeon in Amsterdam)

fold, a para-umbilical defect of the entire abdominal wall will be the result. According to Duhamel[2] this could be the explanation for the occurrence of gastroschisis.

In omphalocoeles the covering layer is not resorbed, because this consists of extra-embryonic mesenchyme, which is identical with Wharton's jelly of the umbilical cord.

Why it should nearly always be the *right* lateral fold which is thus affected remains unexplained by this theory.

Shaw[3] postulates that there is no need to look for a rather elaborate, embryological explanation for the pathogenesis of gastroschisis. The characteristic features of gastroschisis, such as the normal location of the umbilical cord on the abdominal wall (whilst in omphalocoeles the cord usually emerges from the top of the coele), the rarity of herniation of any portion of the liver and the infrequent association with other congenital anomalies, all tend to show that gastroschisis simply amounts to a hernia of the umbilical cord which has ruptured. This rupture takes place *after* completion of the doubling back of the abdominal wall, but *before* the closure of the umbilical ring. The remnants of an existing sac will then have been resorbed *in utero*, leaving the 'clean' edges of the defect, which distinguish gastroschisis so clearly from a ruptured omphalocoele. As the rupture of the umbilical membrane takes place after completion of the morphogenetic phase and given that this is not necessarily

due to a teratogenic action, gastroschisis is rarely accompanied by other than gastrointestinal anomalies.

Shaw also explains why the rupture of this membrane always takes place to the right of the umbilicus. In an embryo of 7 mm, a right *and* a left umbilical vein are present. At a later stage the right umbilical vein disappears and the midgut protrudes into the umbilical cord. Therefore the right side of this protrusion is not as well supported as the left side, where the umbilical cord is concerned. Consequently, a hernia of the umbilical cord would be more likely to rupture on the right side.

## AETIOLOGY

It is generally accepted that teratogenic action before the sixth week of pregnancy may result in omphalocoele. Duhamel reasons that gastroschisis is the result of such an action at a later stage of gestation, between the sixth and eighth week. Such a teratogenic action can also be 'inborn'. This means that the initial 'blueprint' of the child harbours a 'genetic aberration', which causes an abnormal fetal morphogenesis from the very beginning. Consequently, the resultant abnormality is often accompanied by other congenital anomalies. If, on the contrary, the teratogenic action takes place at a later stage – during the organogenetic phase – the resultant anomaly (gastroschisis) will hardly ever be accompanied by other than local defects, such as atresia of the bowel, 'apple peel bowel' and non-rotation of the gut.

Exogenous teratogenic actions in early pregnancy are a well known cause of serious multiple congenital malformations. The tragic consequences of rubella, toxoplasmosis and other infectious diseases, as well as some drugs, have often been described. No one will ever forget the terrible 'thalidomide affair'. Thomas and Atwell[4] furnish no explanation for the origin of omphalocoele, but they do supply clinical reasons for their view that gastroschisis is the end result of an *in utero* rupture of an incarcerated hernia of the umbilical cord. This view is substantiated by the absence of a muscular layer between the defect and the umbilical cord. The common mesentery has led Thomas and Atwell to draw the conclusion that the gut does not herniate through the defect, but is denied the opportunity to return to the abdomen to undergo normal rotation. This theory is supported by the high incidence of gastroschisis cases that are accompanied by atresia of the bowel, probably resulting from the incarceration.

Genetic causes are also mentioned in the literature, especially in connection with omphalocoele. Osuna and Lindham[5] reported four cases of omphalocoele in two generations of one family. One of these children was also suffering from Down's syndrome. Kučera and Goetz[6] described a family with four consecutive children, all omphalocoele patients. Irving[7] described the association of omphalocoele with the Beckwith–Wiedemann syndrome. A brother and sister, born consecutively and both with omphalocoele, were reported by Rott and Truckenbrodt[8]. There are numerous other publications that link omphalocoele with genetics. Contrary to this, Mahour[9] states that 'chromosomal abnormalities do not seem to play an important role in the cause of omphalocoele'. It will be clear from these divergent viewpoints that

neither the cause nor the origin of both omphalocoele and gastroschisis are clearly understood and that the various attempts at explanation are not always convincing.

## TERATOGENIC ACTION OR GENETIC CAUSE?

Particularly regarding omphalocoeles there are arguments in favour of both 'external' and 'internal' teratogenic agents. Experiments have demonstrated that omphalocoeles can be induced in rats by hypoxaemia (Ingalls *et al.*[10]) and with various chemicals. The foregoing supplied ample evidence for the theory that 'internal' factors play a role in the origin of omphalocoele. This evidence is not found in the case of gastroschisis, although a genetic influence has been mentioned on occasion in the literature (Tan[11]).

The importance of determining whether genetic factors may cause omphalocoele as well as gastroschisis, or just omphalocoele, must be obvious. Parents will want to know the risk factor for future pregnancies. Noordijk and Bloemsma-Jonkman[12] studied a series of 51 cases of gastroschisis and large omphalocoeles in order to determine the familial incidence of congenital anomalies, if any. Of this series 37 cases were classified as omphalocoele and 14 as gastroschisis. The result demonstrated that the incidence of congenital anomalies was considerably higher in the families of omphalocoele patients than in the families of patients with gastroschisis. Of these 37 cases of omphalocoele, 15 families emerged with concomitant congenital anomalies and this concerned first and second degree of kinship only. In 3 cases, congenital anomalies had occurred on both the father's and mother's sides. Among the omphalocoeles there was a pair of conjoined twins.

Where the 14 cases of gastroschisis were concerned, congenital anomalies were found in the history of only 1 family. This series was extended at a subsequent date to include 7 additional cases, in none of which concomitant congenital anomalies could be found in the family history. This difference in familial incidence is significant and Noordijik and Bloemsma-Jonkman conclude that genetic factors must play a part in the origin of omphalocoele, while this does not happen in gastroschisis.

Consequently, as regards the risk of other congenital anomalies in future pregnancies, caution is indicated for parents having a child with omphalocoele, but there seems to be no contraindication for other pregnancies in the case of gastroschisis.

## MANAGEMENT

As recently as 10 years ago the survival of a child born with gastroschisis or a ruptured omphalocoele was something of an exception. Today the mortality for gastroschisis has dropped below 25 %. Unfortunately the improvement in the mortality rate of ruptured omphalocoeles has been far less spectacular. In fact this holds true for the entire group of omphalocoeles. There is obviously no question as to the need for surgical intervention in the treatment of gastroschisis and ruptured coeles. Conservative treatment is only feasible in covered coeles.

## Preoperative care

The hazards to which these unfortunate children are exposed are manifold. The major problems can be classified as *anatomical, surgico-technical, microbiological, respiratory* and *nutritional.*

Let us accompany such a child on his hazardous way. Directly upon birth sterile dressings should be applied with a care comparative to surgical sterility to prevent contamination. These dressings also help to check hypothermia, which may become severe during transport if the child is born outside a hospital. These children should be transported in a well heated incubator to a paediatric surgical department. Ventilatory support and suction should be available in the ambulance.

In the hospital the operating theatre should be prepared for emergency. Instant laboratory data should be obtained and an intravenous drip must be inserted. Systemic, broad-spectrum antibiotic therapy must be started immediately, albeit not before swabs are taken from the mass of protruding viscera. As a matter of course the theatre and operating table should be properly heated and the temperature of the child must be monitored continuously.

## Surgical management

The surgeon must check the bowel carefully for continuity and non- or malrotation. The 'checklist' should also include other viscera, such as gallbladder and possibly bile ducts, both kidneys and the diaphragm. If necessary the opening should be enlarged upwards and downwards. This is usually the case in gastroschisis.

The operation of choice entails: *primary closure* of the abdominal wall defect. This is nearly always possible in cases of small or medium sized coeles with a diameter of less than 6 cm, and it may often be possible in gastroschisis. Primary closure may be hazardous or even impossible in cases of large coeles, with a protruding liver, or in antenatal forms of gastroschisis, with congested, rigid and oedematous eviscerated bowel covered by a gelatinous matrix. As a large part of the intestine has developed outside the abdominal cavity, this cavity is 'underdeveloped' and usually too small to accommodate the mass of swollen viscera. Manual stretching of the abdominal wall may help to some extent, but in the majority of cases this will not suffice. The liver often protrudes at least in part and relocating this in its 'proper' place may result in compressing the inferior caval vein or in 'kinking' the hepatic veins. This causes congestion of blood in the lower half of the body, often with fatal results. Desperate attempts at closure will push the diaphragm up too high and impair respiration. Consequently, it is not surprising that surgeons all over the world have looked for safer ways to manage gastroschisis and large ruptured omphalocoeles.

A major breakthrough was the introduction by Schuster[13] of a silastic pouch (Figure 7.5). Using such a pouch, the surplus intestine can be left outside the abdomen. It acts as a kind of artificial coele-wall. At one end the pouch is stitched to the abdominal wall, edge meeting edge of the defect. A

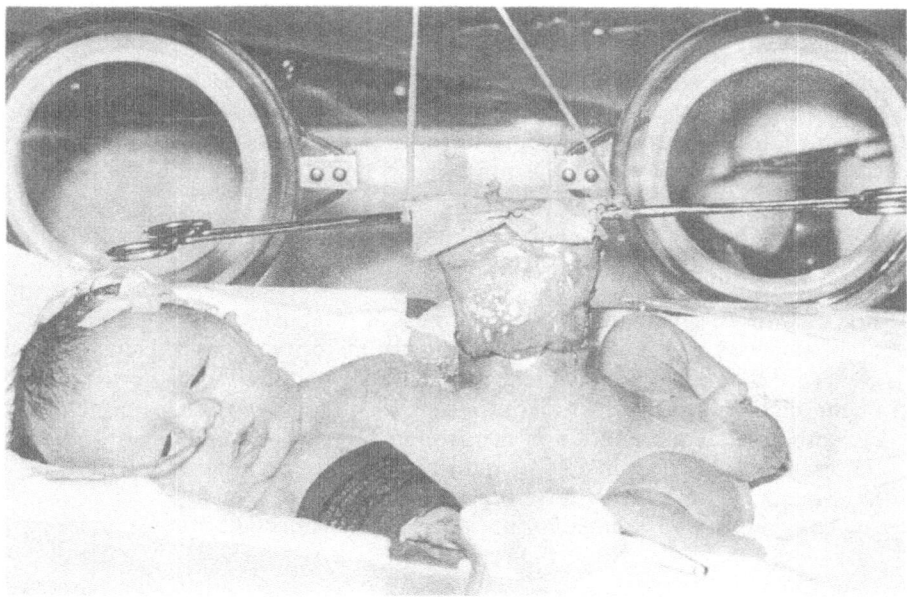

**Figure 7.5** A silastic pouch makes an adequate temporary receptacle, when primary closure is hazardous

clamp at the other end of the pouch facilitates rolling like a tube of toothpaste, which pushes the intestines gently into the abdominal cavity. This should be effected within 10–12 d, because the stitches connecting the pouch to the abdominal wall will start to work loose by then. Using this technique, respiration and circulation remain unimpaired and on the whole the congestion and oedema of the intestines will disappear in 7–10 d. This method works well, especially in gastroschisis where the liver is not protruding and where oedema of the bowel is the main problem. What is more, the defect of the abdominal wall in gastroschisis is usually small and repair is possible without increasing the intra-abdominal pressure too much. The initial technique has been modified by others (Biemann-Othersen[14], Shermeta and Haller[15]) and in this way it is possible to tide over the first critical days of life. In ruptured omphalocoeles however, reconstruction remains a problem, even when a pouch is applied, and many investigators have been looking for more satisfactory and safer procedures.

Willital[16] used 'fluid conserved' dura and found that this takes 3–4 months to be changed into body-own fascia-like tissue. Gharib[17] covered the protruding viscera with amniotic membrane. He reports 8 survivals in 20 cases with this technique. In cases where the silastic pouch worked loose before abdominal closure could be effected, Seashore et al.[18] protected the viscera with porcine skin and amniotic membrane. In their hands these so-called "biological dressings" proved effective. There was only 1 death out of 11 gastroschisis patients treated in this way, while 2 out of 5 omphalocoele patients succumbed.

Gastroschisis patients with a congested mass of bowel plus atresia present a special problem. A primary anastomosis of the oedematous and brittle bowel is hazardous. A relatively safe procedure is to cover the viscera with a pouch in the usual way and *postpone* the anastomosis until the swelling has subsided and the pouch will be removed.

## Non-surgical management

It goes without saying that a 'routine' must be established for large, covered omphalocoeles which can be followed in the majority of cases. Here again, great care should be taken to prevent bacterial contamination as the infection might migrate through the wall of the sac. We do not advocate 'painting' the sac with an aqueous solution of merbromin Mercurochrome (Grob[19]) even though this does result in a good and strong eschar, because several cases of fatal mercury intoxication have been reported.

The coele may be 'painted' or dressed with a disinfectant, but we prefer to

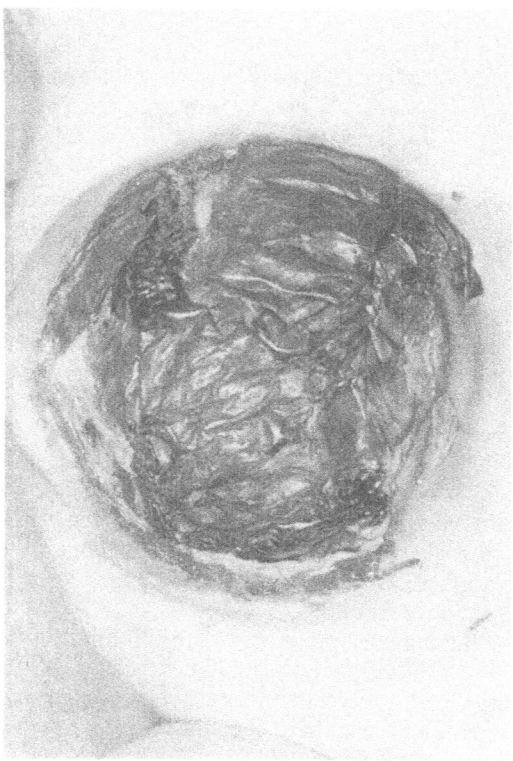

**Figure 7.6** Epithelialization takes place from the edges of the coele

powder it lightly with Nebacetin. Epithelialization will soon start from the edges inward and the coele should be covered in 4–8 weeks. The ensuing eschar will contract slowly and will push the viscera gently back into the abdominal cavity, without impairing respiration or causing circulatory problems. The remaining large hernia may be dealt with electively at a later date (Figures 7.6, 7.7). Many authors consider this a safer procedure than attempts at primary closure, or using techniques based on 'pouch' procedures described above to accelerate the healing process. Ein and Shanding[20] introduced a polymer membrane for the non-operative treatment of large omphalocoeles.

## Postoperative care

The postoperative care in the treatment of omphalocoele and gastroschisis is essentially the same as in any major surgery of neonates. Small and medium sized coeles, in which primary closure has been effected, do not require special measures. One should always be on the alert, however, especially when the patient fails to thrive, for other concomitant anomalies which are not always obvious.

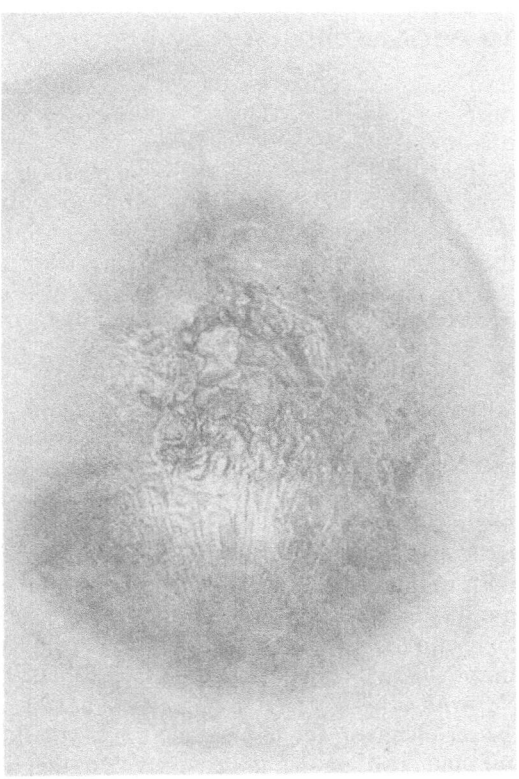

**Figure 7.7**  After total epithelialization a large hernia of the ventral wall remains

Ventilatory support might be necessary during the first few days. In large omphalocoeles and in gastroschisis, this support is needed more often than not for a prolonged period, especially for neonates with a low birth weight. Surprisingly enough, children with a normal birth weight often manage quite well on a Schuster pouch without any respiratory aid. As a matter of course, blood gas samples should be taken at regular intervals during the first 24 h. Sepsis monitoring is essential in all gastroschisis patients as well as children with a ruptured omphalocoele. The majority of these patients develop bacteriologically substantiated septicaemia.

A major problem, especially in gastroschisis, is the prolonged intestinal malfunction which can last several months. Regaining of peristaltic activity is delayed in all these children, and studies by O'Neill and Grosfeld[21] have also shown malabsorption as well as a protein and fat deficiency lasting several months. A gastric tube must be inserted and an intravenous drip should be applied to control fluid and electrolyte losses. Total parenteral nutrition must be started after 7–10 d. Instead of a gastric tube, a gastrostomy might be performed, but the advantages offered hardly warrant the additional risk.

In promising cases, oral feeding may be started after 1 week, but even then it usually takes from 2 to 6 weeks before full enteric feeding can be effected.

## RESULTS AND PROGNOSIS

During the past 15 years the results have improved considerably, especially where gastroschisis is concerned. This is mainly due to more efficient perioperative care. Improved transport facilities, quicker and more accurate laboratory data, specialized anaesthesia, early diagnosis and more effective treatment, let alone prevention of septicaemia, all have contributed to lower mortality and morbidity rates in these patients. Appropriate ventilatory assistance and the improvement of surgical techniques have also contributed. About 50–60 % of the children suffering from gastroschisis are born prematurely or immaturely and their birth weights are below 2500 g. Special treatment in neonatal intensive care units has improved the results in this group (Moore[22]). Up to 1965 the mortality in gastroschisis was reported to lie somewhere between 50 and 70 %, while more recent figures vary between 10 and 25 %[23–25]. The mortality percentage for gastroschisis seems to be stabilizing at these figures, because of unavoidable deaths due to total atresia or total necrosis of the bowel. Consequently, only marginal improvement can be expected in the near future.

In contrast to the marked improvement in the results for gastroschisis, the mortality figure for omphalocoeles has dropped from 40 to 30 % only. To a large extent this is due to the accompanying anomalies, which in many cases are not compatible with life. Stringel and Filler[26] reported 5 deaths out of 42 cases (12 %) of omphalocoele with no concomitant anomalies, while 21 out of 37 patients (56 %) with other accompanying anomalies succumbed.

The slightly better results in the last decade are partly due to the factors mentioned above and also to the improvement in the treatment of the concomitant malformations. Advances in neonatal cardiac surgery have benefited several patients with omphalocoele.

# Acknowledgement

The author wishes to thank Alice Goslinga Ribbink, Translator/Stylistic Editor, for her assistance in the preparation of this chapter.

# References

1 Müntener, M. (1970). Zur Genese der Omphalocele und Gastroschisis (paraumbilikale Bauchwanddefekt). *Z. Kinderchir.*, **8**, 380

2 Duhamel, B. (1963). Embryology of exomphalos and allied malformations. *Arch. Dis. Child.*, **38**, 142

3 Shaw, A. (1975). The myth of gastroschisis. *J. Pediatr. Surg.*, **10**, 235

4 Thomas, D. F. M. and Atwell, J. D. (1976). The embryology and surgical management of gastroschisis. *Br. J. Surg.*, **63**, 893

5 Osuna, A. and Lindham, S. (1976). Four cases of omphalocele in two generations of the same family. *Clin. Genet.*, **9**, 354

6 Kučera, J. and Goetz, P. (1971). Exomphalos in four consecutive pregnancies. *Humangenetik*, **13**, 58

7 Irving, I. M. (1976). Exomphalos with macroglossia: A study of eleven cases. *J. Pediatr. Surg.*, **2**, 499

8 Rott, H. D. and Truckenbrodt, H. (1974). Familial occurrence of omphalocele. *Humangenetik*, **24**, 259

9 Mahour, G. H. (1976). Omphalocele-collective review. *Surg. Gyn. Obst.*, **143**, 821

10 Ingalls, T. H., Avis, F. R. *et al.* (1953). Genetic determinants of hypoxia-induced congenital anomalies. *J. Hered.*, **44**, 185.

11 Tan, K. L. (1973). Abnormalities of the umbilical region: a clinical study. *Br. J. Clin. Pract.*, **27**, 323

12 Noordijk, J. A. and Bloemsma-Jonkman, F. (1978). Gastroschisis no myth. *J. Pediatr. Surg.*, **13**, 47

13 Schuster, S. R. (1967). A new method for the staged repair of large omphaloceles. *Surg. Gynec. Obst.*, **125**, 837

14 Biemann-Othersen, H. and Hargest, T. S. (1977). A pneumatic reduction device for gastroschisis and omphalocele. *Surg. Gynec. Obst.*, **144**, 243

15 Shermeta, D. W. and Haller, J. A. (1975). A new preformed transparent silo for the management of gastroschisis. *J. Pediatr. Surg.*, **10**, 973

16 Willital, G. H. (1976). Klinische Erfahrungen mit Dura Implantationen in der Neugeborenenchirurgie. *Z. Kinderchir.*, **19**, 16

17 Gharib, M. (1975). Versorgung der pränatal rupturierten Omphalocele und das para-umbilikale Bauchwanddefektes mit geburtseigenen Eihäuten. *Münch. Med. Wochenschr.*, **117**, 1555.

18 Seashore, J. H., MacNaughton, R. J. and Talbert, J. L. (1975). Treatment of gastroschisis and omphalocele with biological dressings. *J. Pediatr. Surg.*, **10**, 9

19 Grob, M., (1965). Conservative treatment of exomphalos. *Arch. Dis. Child.*, **38**, 148

20 Ein, S. H. and Shanding, B. (1978). A new nonoperative treatment of large omphaloceles with a polymer membrane. *J. Pediatr. Surg.*, **13**, 255

21 O'Neill, J. A. and Grosfeld, J. L. (1974). Intestinal malfunction after antenatal exposure of viscera. *Am. J. Surg.*, **127**, 129

22 Moore, T. C. (1977). Gastroschisis and omphalocele: clinical differences. *Surgery*, **82**, 561

23 Rickham, P. P. and Johnston, J. H. (1969). Exomphalos and gastroschisis. *J. Neonatal Surgery*, p. 254. (London: Butterworths)

24 Hollabaugh, R. S. and Boles, E. T. (1973). The management of gastroschisis. *J. Pediatr. Surg.*, **8**, 263

25 Lewis, J. E., Kraeger, R. K. and Danis, R. K. (1973). Gastroschisis, ten year review. *Arch. Surg.*, **107**, 218

26 Stringel, G. and Filler, R. M. (1979). Prognostic factors in omphalocele and gastroschisis. *J. Pediatr. Surg.*, **14**, 515

# 8
# Urethral duplication

M. GHARIB

---

The first observation of a double urethra is probably ascribable to Aristotle[1]. More precise indications are contained in publications of the sixteenth and seventeenth centuries by Donatus (1586) and Hildanus (1646)[2]. For descriptions of cases from early autopsies we are indebted to the work of Schatz[3].

Compared to other genital malformations, urethral duplications are rare. Adair[4] estimates the frequency of occurrence of double penis and urethra as roughly 1 in 5.5 million livebirths. In the literature the data are largely uncertain concerning the percentual incidence of isolated urethral duplication. The abnormality often remains undetected or is faultily diagnosed. It is therefore doubtful whether any statistics on frequency give a true picture.

Moreover, the data regarding number of individual observations, as collated by various authors, are at variance (Table 8.1).

All authors agree, however, that urethral duplication occurs more frequently in males than in females[1,5−7]. This is explicable by the fact that in males a more elaborate differentiation of the urethral anlage takes place beyond a certain phase of embryonic development[2,8,9].

## EMBRYOLOGY

In both sexes, the first phase of embryonic development of the outer genital organs, including the urethra, is via the indifferent urogenital sinus that derives from the cloaca. The cloaca, a common cavity for the intestinal and urogenital systems and which initially has no outlet, due to a cloacal membrane, extends from navel to rump; this is of importance for comprehending the causes of epispadias and exstrophy of the bladder.

In the 3rd embryonal week, mesenchymal cells wandering bilaterally from the region of the primitive streak to the vicinity of the cloacal membrane form the two cloacal folds. The urorectal septum, growing in a frontal plane, divides the cloaca into the anterior urogenital sinus and the posterior rectal anlage.

When the urorectal septum reaches the cloacal membrane this latter becomes subdivided into the anal and urogenital membranes. Concomitantly, the cloacal folds become partitioned into an anterior section, the urogenital

**Table 8.1** Varying data given in the literature concerning total number of individual observations

| Author | Year | Complete duplication | Total |
|---|---|---|---|
| Gross and Moore[1] | 1950 | 19 | 83 |
| Wrenn and Michie[32] | 1957 | 23 | % |
| Boissonnat[5] | 1961 | 23 | % |
| Casselman and Williams[11] | 1966 | 30 | > 100 |
| Montagnani[6] | 1968 | 27 | 88 |
| Tripathi and Dick[33] | 1969 | 27 | % |
| Karanjavala[34] | 1970 | 25 | % |
| Schirmer[17] | 1970 | 31 | 202 |
| Wilson[19] | 1971 | 44 | % |
| Singh[35] | 1973 | 40 | % |
| Redman[36] | 1975 | 45 | % |

folds, and a posterior section, the anal folds. In the embryo of 18 mm crown–rump (C–R) length the anal membrane ruptures first, the urogenital membrane rupturing shortly afterwards. The urogenital folds, distinctly observable in the 4th embryonal week, grow more extensive. In the 5th week of embryonic development they fuse in the cranial midline forming the genital tubercle. At the same time, two eminences appear on either side of the urogenital folds. These are known as the genital or labioscrotal swellings. With the rupture of the urogenital membrane and formation of the genital tubercle, the primitive urogenital orifice is formed. Up to this stage there is outwardly no means of distinguishing male from female (Figure 8.1).

Differences first become apparent in 4 mm long embryos[8,9]. The urogenital folds are implicated in outgrowth of the genital tubercle ventrally, whereby the urogenital orifice is drawn out to form the urethral groove. The changes subsequent to this indifferent stage are in the female relatively slight: the genital tubercle becomes first the phallus and then the clitoris; the urogenital folds become the labia minora. The space between the folds, i.e. the region of the so-called urethral groove, forms the vestibulum vaginae and the genital swellings become the labia majora[8,10].

In the male, the degree of morphological differentiation surpasses that of the female. The genital tubercle, after becoming the indifferent phallus, grows much more extensively into the penis, at the undersurface of which the urogenital or urethral folds fuse together over the urethral groove making this a tube, the walls of which become the pars spongiosa of the urethra. Only the distal portion of the urethra in the region of the glans penis forms by ectodermal ingrowth and later canalization, the definitive ostium urethrae forming at the tip of the glans[2,10]. The genital swellings unite medially to become the scrotum (Table 8.2).

## THEORIES ON GENESIS OF URETHRAL DUPLICATION

The precise mechanism triggering formation of a double urethra has not yet been elucidated. Theories thus far advanced on the aetiology of this malformation can be summed up as follows[5,11,12].

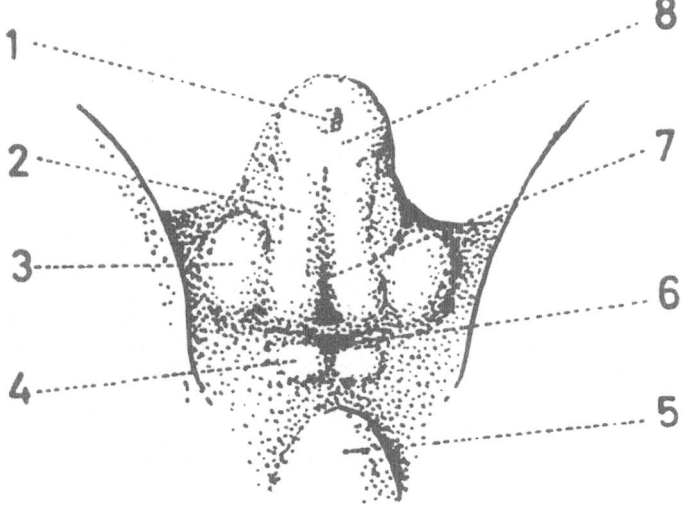

**Figure 8.1**  External genitalia of embryo of 18 mm C–R length in indifferent stage. 1, Epithelial tubercle; 2, urogenital fold; 3, genital swelling; 4, anal tubercle; 5, tail in process of involution; 6, anus following rupture of anal membrane; 7, sulcus urogenitalis prior to rupture of the urogenital membrane – between anus and urogenital membrane can be seen the primitive perineum; 8, genital tubercle (phallus) (*from* Pompino *et al.*[8])

**Table 8.2  Development of external genital organs**

| Male | Indifferent primordia | Female |
|------|----------------------|--------|
| Corpora cavernosa penis | Genital tubercle | Clitoris |
| Termination of corpora cavernosa penis | Glans of phallus | Glans clitoridis |
| Corpus spongiosum and glans penis | Urogenital (urethral) folds | Labia minora |
| Scrotum | Genital (labioscrotal) swellings | Labia majora |
| Pars spongiosum urethrae | Urogenital sinus | Vestibulum vaginae |

(1)  Division of the urethral anlage due to an abnormal continuation of the dividing process through the urorectal septum.

(2)  Retarded fusion of the paired buds of the genital tubercle, this fusion then extending too far dorsally.

(3)  Division of the urethral groove.

(4)  Imperforate cloacal membrane, the mesoderm being prevented from strengthening the mid-section of the abdominal wall below the navel; further consequences of this can be exstrophy of the bladder and, along with urethral duplication, the occurrence of a dorsal chorda or fibrous band.

Another theory given consideration by Selvaggi and co-workers[12] is that an inflammatory process occurs leading to distal occlusion of the normal urethra and causing a fistula or canalized diverticulum to appear.

There is, however, no comprehensive hypothesis to account for all possible forms of urethral duplication. A few theories offering a plausible explanation for particular variations have found general acceptance.

## CLASSIFICATION AND DEFINITION

The ducts that more usually lie dorsal to, but which in rare cases can lie ventral to, the normal urethra, are regarded as accessory urethrae. These can grow to varying length; they can communicate with the normal urethra, egress as fistulas or end blindly in the viscera. Complete duplication occurs predominantly in conjunction with bladder and genital duplications. So far, numerous variants of urethra doubling have been reported in the literature consequent on clinical examinations. Many attempts have been made to classify the different forms of this malformation, the first ascribable to Taruffi (1891)[29]. For practical purposes we have found the following classification the most useful.

(1) Urethral duplication with double formation of colon or complete caudal double–malformations (Figure 8.2).

(2) Urethral duplication with double bladder and duplicated outer genital organs (Figure 8.3).

(3) Urethral duplication with single bladder and single genital anlage. This group is subdivided as follows:

    (i) functionally complete double urethra or duplicate urethrae that intercommunicate (Figure 8.4);

    (ii) functionally incomplete double urethra or duplicates that end blindly; here again, two groups can be distinguished, those with proximal blind endings and those with distal blind endings (Figure 8.5).

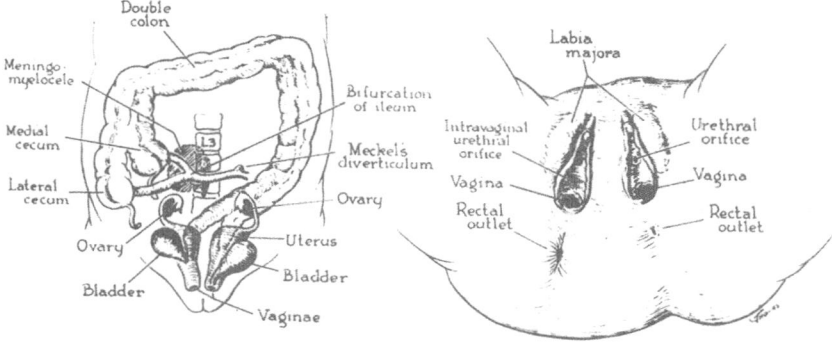

**Figure 8.2** Bilateral duplication of the urethra, bladder, external genitalia, and colon. (*From* Gray and Skandalakis[2])

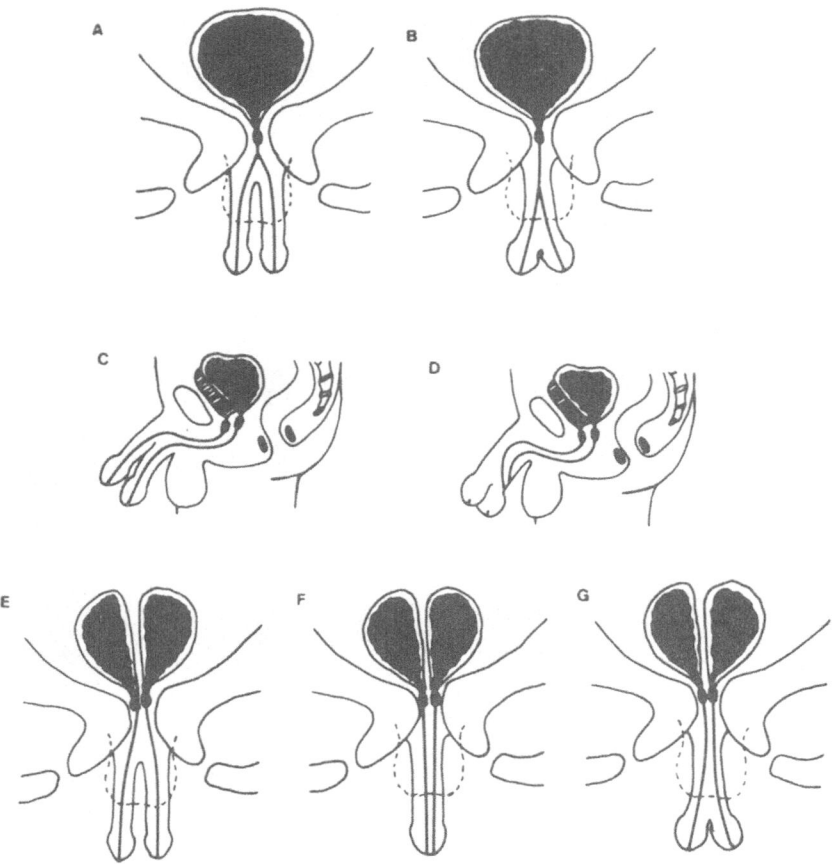

**Figure 8.3** Different variants of complete duplication of bladder and urethra

Among the functionally complete urethral duplications we count such forms as those where there is excretion of urine from both urethral orifices no matter whether the ducts are fully developed in length, with separate egress from the bladder, or whether there are proximal or distal intercommunications.

Thus, there occur some forms of complete urethral duplication with isolated egress from the bladder and two separate orifices at the tip of the glans (Figure 8.4a).

On the other hand, variations exist where the urethra, either in its proximal or in its distal section, is only partially duplicated. Hence, the two urethrae can either leave the bladder separately and become conjoined to form a common urethra in the distal region (Figure 8.4b), or they can start off as one urethra and divide into two canals in the distal section (Figures 8.4, c and 8.4, d). The secondary urethra may have an outlet either ventrally or dorsally in the penis shaft, or the opening may be at the base of the penis or in the region of the perineum (Figures 8.4, c and 8.4, d).

149

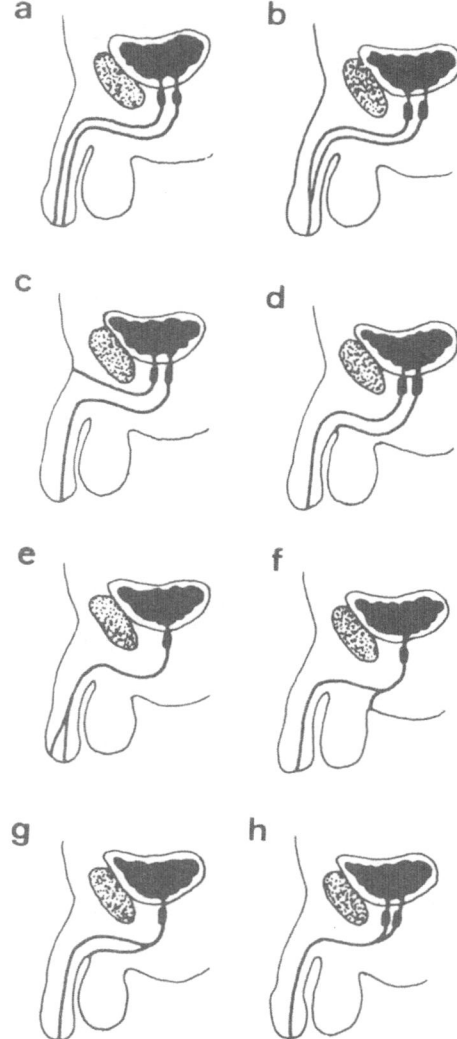

**Figure 8.4** Variants of functionally complete urethral duplication

We understand by functionally incomplete urethral duplication such forms as those in which there is excretion of urine from one of the two ducts while the second urethra ends blindly, either proximally or distally. Blindly ending urethrae can be of varying length and have a varying topographical position. For example, the accessory urethra can be almost fully developed and end blindly just short of the bladder (Figure 8.5, 1). Alternatively, it can stretch for only a few millimetres or centimetres and be situated in the glans penis (Figure 8.5, 4). Very often, the outer orifice is in the region of the dorsal penis base whereby the accessory urethra runs in the direction of the neck of the bladder

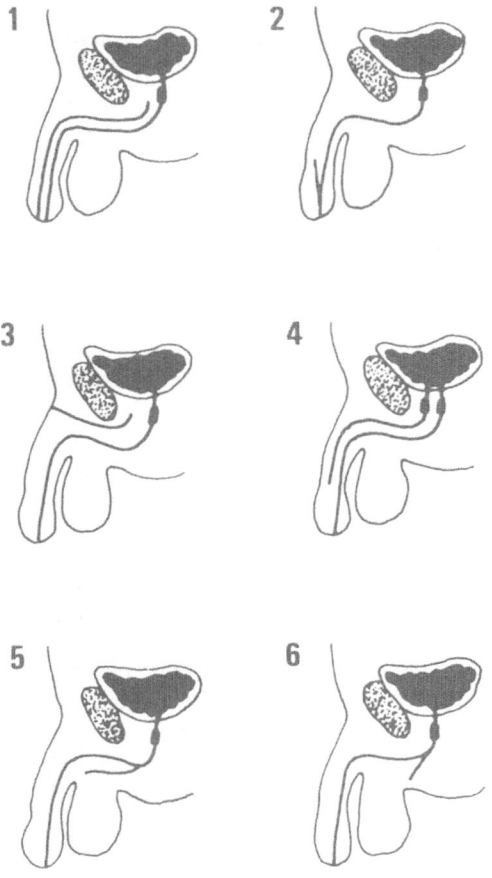

**Figure 8.5**  Variants of functionally incomplete urethral duplication

for a few centimetres, then ends blindly (Figure 8.5, 3). The duplicate urethra can also run in the direction of the ventral penis shaft at the penis base and there have an outlet, or emerge at the perineum (Figures 8.5, 5 and 8.5, 6).

## SYMPTOMATOLOGY AND DIAGNOSIS

Observations of clinical manifestations of urethral duplication depend on the localization and the connection to the bladder, as well as on the presence of additional malformations. Manifestations ranging from urethrae with asymptomatic course to serious infravesicular miction disturbances can be rendered more complicated by recidivate urinary infections or enuresis and incontinence (Table 8.3).

Where a complete diphallus is present, the possibility of a double urethra can be assumed and is not difficult to check on. But in cases of associated

F

**Table 8.3  Subjective and objective indications
of urethral duplication**

1 Visible opening
2 Effluence of secretion or pus
3 Dribbling of urine
4 Urinary infections
5 Incontinence
6 Micturition disturbances
7 Pain on miction
8 Local pains

malformations such as exstrophy of the bladder or vesico-intestinal fissure with caudal duplication, abnormalities in the outer genital organs connected with these may be overlooked or not be recognized as such (Figure 8.6).

Urethral duplication is either externally observable in the form of two urethral openings or obscured by absence of a secondary orifice in the genital region. If, during micturition, two jets emerge from two separate openings, the diagnosis is clear even in very young children, the double outlet being revealed on inspection. Quite often the parents themselves notice an abnormal outlet in the region of the penis base and report this. Schrimer[17] collated 202 case histories of urethral duplication and established the occurrence in 88 infants (43 %) of a blindly ending urethral duplicate with external outlet in the dorsal region of the penis; in some cases, a pus-like secretion exuded from this opening. Inflammation in the secondary canal can cause topical pain. Not infrequently a second circular outlet is found, on closer inspection, a few millimetres above, or more seldom a few millimetres below, the proper meatus externus. The accessory urethral outlet at the glans is often stenotic, especially

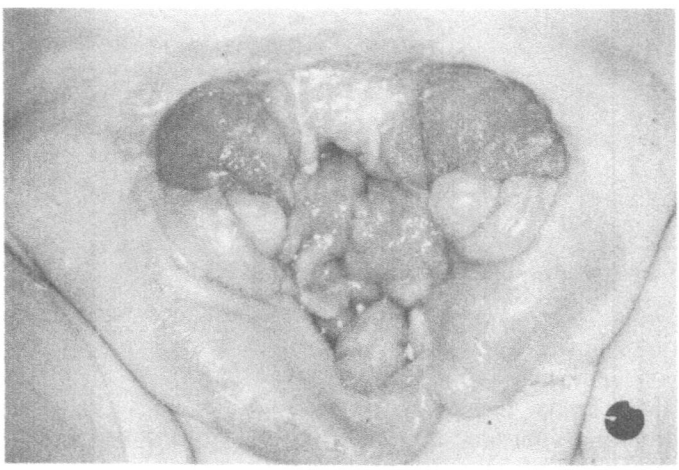

**Figure 8.6**  External aspect of child with caudal double-malformation

as a consequence of an infection[18]. When the opening is situated on the dorsal surface of the penis it is often slit-like, resembling an epispadias; it may indeed be faultily diagnosed as such, particularly if the outlet of the urethra proper on the ventral side of the sulcus coronarius is difficult to discern[11]. Due to the chorda or fibrous band stretching from the end of the dorsally situated urethra to the glans, the penis often has an epispadiac aspect[19]. Frequently, patients in early infancy with urethra doubling are brought to the paediatric-urological clinic for treatment because of recidivate urogenital infections. The root cause of such infections is usually an urodynamic micturition disturbance[1,12,18].

Less often, the urethral duplication can be the cause of incontinence if the accessory urethra is complete and leaves the bladder outside the sphincter-muscle area. There is then, in addition to normal miction, an uncontrolled and fairly strong excretion of urine from the secondary urethral opening[1,7,18].

Occasionally, it can lead to nocturnal enuresis and incontinence through stress by reason of a deficient or impaired function of the sphincter-muscle, as a consequence of both urethrae leaving the bladder within the region of this muscle[20].

Urodynamic micturition disturbances are not infrequently due to an urethral double-malformation giving rise to the known symptoms of an infravesicular obstruction. Miction disturbances may result from a circular stenosis at the external orifice, or from short or long urethral stricture, particularly in the bulbar region. Occasionally, the urethra is underdeveloped throughout its entire length and has the form of a thin, thread-like tube. In some cases prestenotic outpocketings or diverticulum formations are ascertainable which, during miction, become tautly filled, thereby compressing the ventral urethral anlage with egress in the perineum. This can bring about grave excretory disturbances with retrograde flow into the seminal tubules; furthermore it can give rise to sacculated bladder, fasciculated bladder and vesico-ureteral reflux (Figure 8.7)[21,22].

Urethrae with external blind endings usually are clinically inconspicuous and remain undetected unless an infection with accompanying pain and urethritis, resulting from blocked urine flow or static urine in the blind channel, prompts a thorough internal examination. For diagnosing such conditions a miction-cysto-urethrogram (MCU) is primarily indicated. Inspection of the two urethral orifices, if present, with catheters and probes, can be rendered difficult by stenotic conditions. Where it is not possible to insert a catheter, opacifying fluid must be administered by suprapubic injection.

Instillation of dye solutions can sometimes provide evidence of an interconnection between the two urethrae. Cysto-urethroscopic examinations can help reveal where the secondary urethra inosculates with the urethra proper or where it egresses from the bladder, or can provide information concerning possible ectopic inlets of the ureter into the cervix of the bladder, which is often an accompanying feature (Figure 8.8).

Further diagnostical aids include radiological examinations of the upper urinary passages, vertebral column and colon, in order to exclude the possibility of concomitant intestinal and spinal malformations, or some other

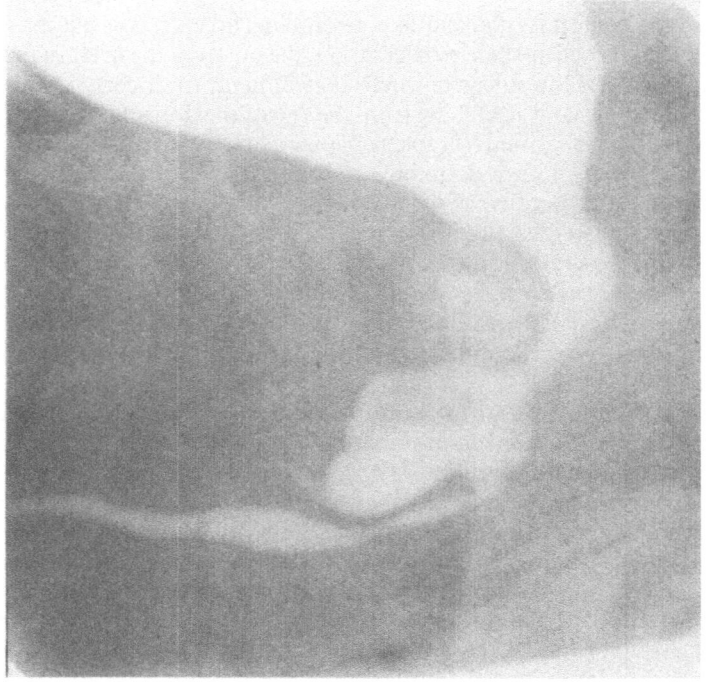

**Figure 8.7** Tautly filled prestenotic section of dorsal urethra with compression of ventral urethra, leading to an infravesicular excretory disturbance

maldevelopment in the region of the kidneys and urinary ducts (entero-urovertebral syndrome)[23-25].

## THERAPY

For treatment of urethral duplication, various therapeutic possibilities are recommended.

## Excision

Contrary to the opinion expressed by some authors, we agree with Gross and Moore that in cases of incomplete urethral duplicates, extirpation is indicated when the supernumerary canal is subject to recidivate infections, and also from a cosmetic standpoint[1,26].

*Clinical case history (B. E., born 29 July 1974)*
Near the root of the penis of this barely 2-month-old, otherwise normal, male infant could be observed, on the dorsal penile surface, a small cutaneous orifice that was palpable with a probe to a depth of 2 cm. No connection to the

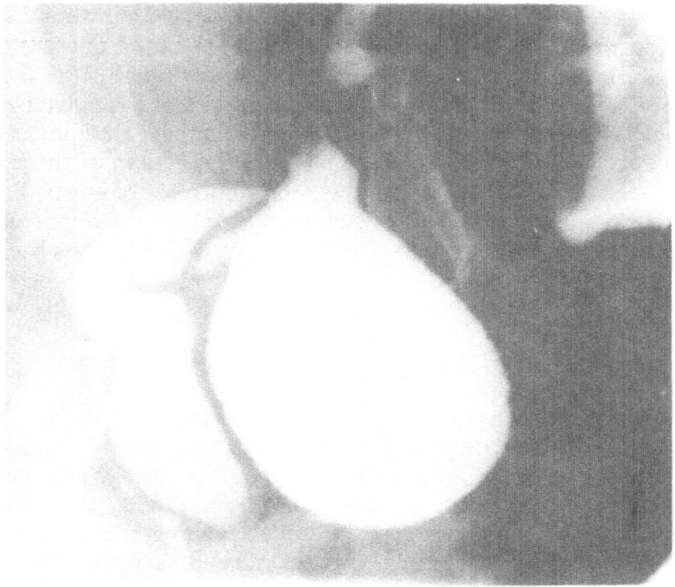

**Figure 8.8** Ectopic inlet of ureter in case of urethral duplication with vesico-ureteral reflux

bladder was established. Surgery was performed on 24 September 1974. Operative procedure: total extirpation of the 3 cm long canal. Histological examination of the blindly ending duct revealed that this was an incomplete urethral duplication (Figure 8.9).

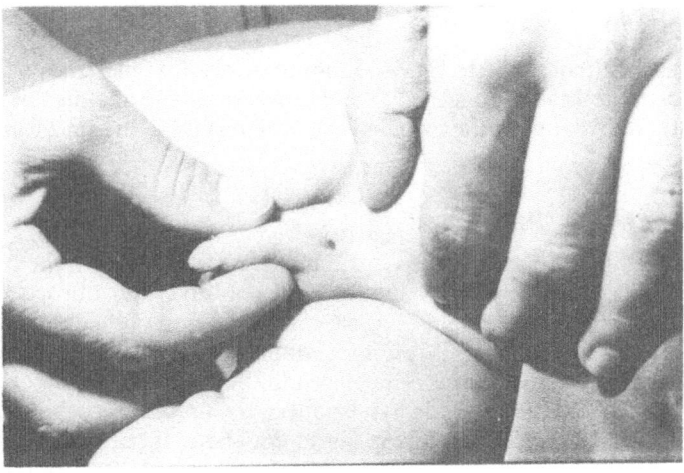

**Figure 8.9** Orifice of incomplete duplicate urethra at dorsal base of penis

Where the accessory urethral duct is complete, total excision is the most successful form of therapy, provided topographical conditions permit this without undue risk. Total excision is particularly indicated if the accessory urethra egresses from the bladder outside the area of the sphincter-muscle, again provided that operative removal does not endanger the sphincter. If a chorda or fibrous band is present, leading from the ostium of the spurious urethra to the glans penis and giving the penis an aspect of epispadias, it is advisable that during resection of the accessory urethra the opportunity be taken of removing this too; the band can cause erection difficulties because of its inelasticity[1,21,27].

## Treatment of urinary infections

If surgery for maintaining continence involves too great a risk, then therapy must be confined to medication. This is especially so in cases of both complete and incomplete urethral duplication which have led to infections of the urinary passages in very young children. Until such time as the anatomical conditions allow corrective surgery to be undertaken without danger, i.e. at some time beyond the second year of life, the infection should be kept under control by antibiogram-indicated therapy[27].

## Induration

Therapy by obliteration, using substances that induce aseptic inflammation and subsequent hardening, causing the lumen of the accessory urethra to be displaced, is seldom undertaken nowadays. The chances of success by this method are too difficult to assess or predict[1,24].

## Fenestration

Reports appear from time to time describing successful fenestration between two adjacent urethrae. Gross and Moore, however, advise against this method because there is danger of a turbulence of the resulting stream of urine[1,28].

## Urethral reconstruction

When in urethral duplication the ventral urethra has its outlet in the perineum near the anus, and the second urethra is thread-like and fibrotic, or with extenuated stenosis in the bulbar region, reconstruction of the ventral urethra in the sense of a hypospadias is recommendable. In the first operative step, the urethral opening near the anus is displaced in the direction of the scrotum; in the second step, as in high-grade hypospadias, the urethra is reconstructed by means of a cellulocutaneous flap as far as the glans. If the dorsal urethra is found to be of normal calibre in the region of the penis shaft, then the ventral urethra in the perineum region can be mobilized to effect an end-to-side anastomosis between the two urethrae[26,27].

## Clinical case-history (J. I., born 8 February 1975)

The barely 7-month-old, lively male infant was accepted as a resident patient and repeatedly given treatment for recidivate urinary-passage infection. Two urethral openings, one at the glans penis and one near the anus, had already been conspicuous at birth. Exterior examination at the clinic revealed nothing else unusual. An i. v. pyelogram disclosed a single left kidney but no other defect. In the miction-cysto-urethrogram a well-palpable opening was revealed in the middle of the perineal region, through which a catheter could be manipulated as far as the lumen of the bladder. Roentgenography revealed a well-developed prostate gland with retrograde filling of the seminal ducts. Most of the opacifying fluid flowed through the urethra with outlet at the perineum. Part of the fluid exuded through a thread-like canal between the middle of the prostate region and the penis shaft and then through a dilated urethra with outlet at the glans region. This was a case of urethral duplication with a hypoplastic stretch of the normally located passage and a ventrally lying duplicate from which most of the urine was excreted (Figure 8.10).

When the patient had reached the age of 2 years, the perineal urethra with orifice near the anus was mobilized in the perineal region, as a first operative step. A stenotic stretch in the perineum could be demonstrated by inserting a catheter (Charriere 8) into the distal section of the dorsal urethra; the tip of the catheter marked the site of transition between the dilated and stenotic

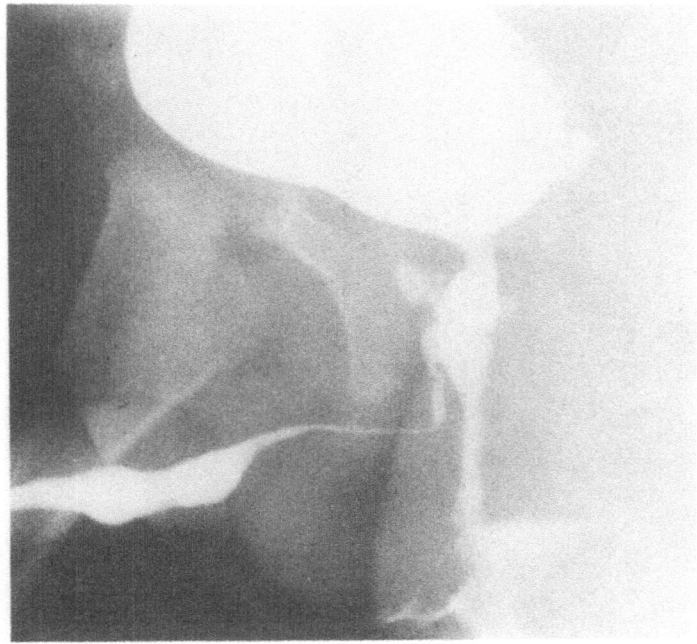

**Figure 8.10** MCU. Demonstration by resting catheter of ventral urethra with external orifice below scrotum in the perineum

portions of the urethra. At this site an end-to-side anastomosis between the ventral and dorsal urethrae was effected.

Postoperatively, a fistula formed in the area; corrective surgery for this was performed in a second operative step (Figure 8.11).

Tests 6 months later showed that the child was able to micturate in an intermissive strong stream (Figure 8.12).

## DISCUSSION

Urethral duplication is, with all its variations, a rare clinical phenomenon. This is substantiated by the small total number of cases described in the literature and by the meagre and conflicting statistical data.

Individual authors have early called attention to the existence of different forms of this abnormality and to the numerous combined variants and classifications[29].

Through paucity of information and experience, different writers have interpreted urethral duplication inconsistently. Furthermore, there is no uniform terminology: double urethra, urethral duplication, accessory urethra, partial urethral duplication, normal and abnormal urethrae.

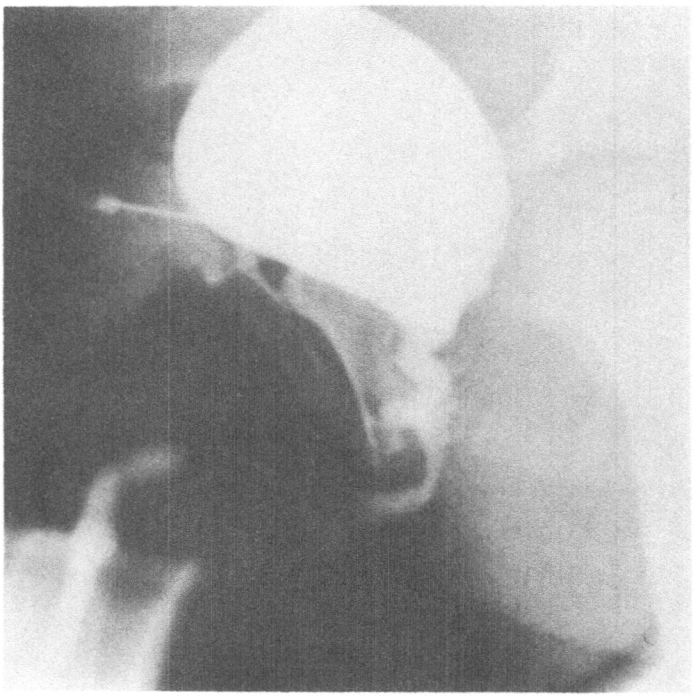

**Figure 8.11** State after transpositioning of ventral urethra and performance of end-to-side anastomosis to bypass the stenotic dorsal urethra

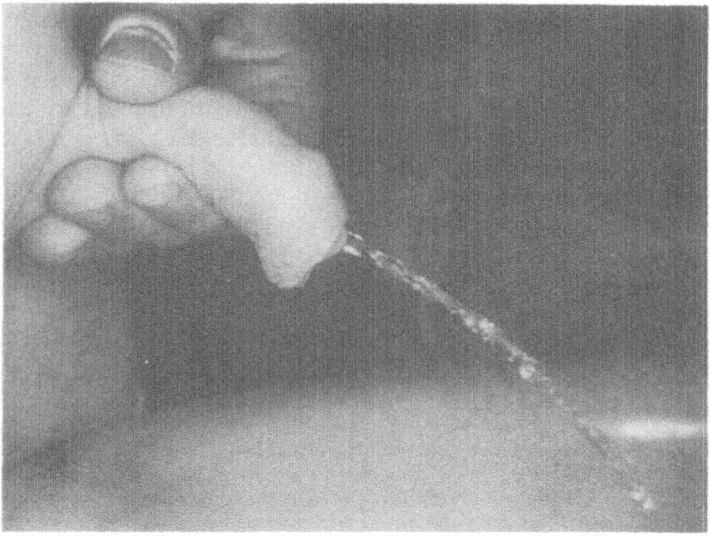

**Figure 8.12** Strong stream of urine following operation

We use the term 'accessory urethra' to mean the functionally non-standard urethra, which as a rule lies dorsal to the normal urethra.

In clinical practice, urethrae are referred to as complete or incomplete according to functional aspects. Although no comprehensive theory covers all localizations and forms of urethral duplication, we agree with Effmann and co-workers[14] that blindly ending urethrae invariably lie dorsal to the functional urethra, irrespective of whether the meatus externus is ectopic (hypospadias) or at the normal site. The dorsally lying urethra, even if exhibiting a normal course through the shaft of the penis to the tip of the glans, frequently has an extensive stenotic stretch in the bulbar section, or is altogether a thread-like, fibrous structure. The ventral urethra very often egresses near the anus in the perineal region. In such cases, Williams[26] advises as a first operative step the displacement of the urethral outlet in the direction of the scrotum, and as a second step urethral reconstruction with cellulocutaneous flaps, as in overt hypospadias. In this way a short, ventral urethral duplicate can be transformed into a complete substitute urethra.

In a similar case, though here with extensive stenosis in the bulbar region, we were able, by mobilizing the ventral urethra with its outlet near the anus, to effect an end-to-side anastomosis, joining it to the dorsally lying urethra and thus bypassing the stenotic section.

Due to rare occurrence and manifold variations of the malformation, there is so far no standardized concept either as regards classification or therapy. But although such abnormalities seldom have clinical significance, their existence must be reckoned with if incontinence and other miction disturbances do not respond to routine therapy for urinary infections and enuresis.

159

Urethral duplication combined with other congenital malformations is not infrequently encountered. Thus, in some cases concomitant vertebral malformations are described, such as wedge vertebrae in the lumbosacral region of the spine and spina bifida. Accompanying malformations of the kidneys and adrenal glands, of the genitals and of the rectoanal region, as well as exstrophy of the bladder, are described in the literature[24,25,30]. Gelbke[31] has drawn attention to the possibility of a pseudo-urethral duplication being of iatrogenic origin, as a consequence of inadvertent lesion to the urethra during proctic surgery.

## References

1 Gross R. E. and Moore, T. C. (1950). Duplication of the urethra. *Arch. Surg.*, **60**, 749

2 Gray, S. W. and Skandalakis, J. E. (1972). *Embryology for Surgeons*, pp. 519–549. (Philadelphia: Saunders)

3 Schatz, D. (1872). Ein besonderer Fall von Missbildungen des weiblichen Urogenitalsystems. *Arch. Gynäkol.*, **3**, 304

4 Adair, E. L. and Lewis, E. L. (1960). Ecotopic scrotum and diphallia. *J. Urol.*, **84**, 115

5 Boissonnat, P. (1961). Two cases of complete double functional urethra with a single bladder. *Br. J. Urol.*, **33**, 453

6 Montagnani, C. A. and Pampaloni, A. (1968). Duplicità completa della vescica e dell'uretra e malformazioni associate. *Minerra Chir.*, **23**, 1143

7 Taguchi, H. and Horiuchi, M. (1967). Duplication of the urethra in a female child. *Jpn. J. Urol.*, **58**, 237

8 Pompino, H. J., Zickgraf, Th., Pietschmann, J. H. and Schmidt, W. (1969). Langzeitkatamnesen von 164 Hypospadien. *Z. Kinderchir.*, **7**, 519

9 Starck, D. (1975). *Embryologie.* (Stuttgart: Georg Thieme)

10 Langman, J. (1970). *Medizinische Embryologie.* (Stuttgart: Georg Thieme)

11 Casselman, J. and Williams D. I. (1966). Duplication of the urethra *Acta Urol. Belg.* **34**, 535

12 Selvaggi, F. P. and Goodwin, W. E. (1972). Incomplete duplication of the male urethra. *Br. J. Urol.*, **44**, 495

13 Bramwit, D. N. and Ziter F. M. (1970). Accessory urethral channel. *Radiology*, **94**, 359

14 Effmann, E. L., Lebowitz, R. L. and Colodny, A. H. (1976) Duplication of the urethra. *Radiology*, **119**, 179

15 Reidy, J. P. (1965) A case of duplication of penile urethra. *Br. J. Plast. Surg.*, **18**, 199

16 Traversa, F. P., Bossu, M. and Della Porta, G. (1969). Una malformazione uretrale rara: L'uretra bifida incompleta. *Minerva Pediatr.*, **21**, 2080

17 Schirmer, U. (1970). Doppelbildungen der Harnröhre. *Z. Kinderchir.*, **8**, 436

18 Müller, H. (1975). Urethraduplikatur. *Z. Kinderchir.*, **17**, 278

19 Wilson, A. N. (1971). Complete dorsal duplication of the male urethra as an isolated deformity presenting as glandular epispadias. *Br. J. Urol.*, **43**, 338

20 Olsen, J. G. (1966). Complete urethral duplication in a boy. *J. Urol.*, **95**, 718

21 Piroth, P., Gharib, M. and Bliesener, J. A. (1976). Komplette und inkomplette Urethraduplikaturen. *Urologe*, **15**, 233

22 Williams, D. I. and Nash, D. F. (1968). *Paediatric Urology.* (London: Butterworths)

23 Heinisch, H. M. and Weidle, I. (1967). Komplette Verdoppelung der männlichen Urethra en eckt im Mictions-urethrogramm *Ann. Radiol.*, **10**, 265

24 Hermann, G. and Goldmann, H. (1973). Double urethra with vertebral anomaly. *Int. Surg.*, **58**, 574

25 Mehan, D. J. and Gonzales, J. H. (1975). Urethral duplication. *Urology*, **6**, 476

26 Williams, D. I. and Bloomberg, S. (1976). Bifid urethra with preanal accessory tract (Y-duplication). *Br. J. Urol.*, **47**, 877

27 Gharib, M. (1980). Systematik, Klinik und Therapie der unbehandelt mit Fertilitätsstörungen einhergehenden Missbildungen der Harnröhre und Penis. *Therapiewoche*, **30**, 320

28 Durrani, K. M., Shah, P. I. and Kakalia, G. R. (1972). Interurethral fenestration for a case of double urethra with hypospadias. *J. Urol.*, **108**, 586

29 Taruffi, C. (1891). Sui canali anomali del pene. *Boll. Sci. Med. Bologna*, **2**, 275

30 Oelsnitz, G. (1977). Kolon-Harnblasen-Urethraduplikatur. *Z. Kinderchir.*, **21**, 381

31 Gelbke, H. (1968). Beseitigung eines iatrogenen Urethradefektes mit urethro-analer Fistel bei recto-analer Missbildung. *Z. Kinderchir.*, **5**, 429

32 Wrenn, E. L. and Michie, A. J. (1957). Complete duplication of the male urethra. *Ann. Surg.*, **145**, 119

33 Tripathi, V. N. and Dick, V. S. (1969). Complete duplication of male urethra. *J. Urol.*, **101**, 866

34 Karanjavala, D. K. (1970). An unusual case of complete reduplication of the urethra. *Aust. N.Z. J. Surg.*, **39**, 284

35 Singh, J. P., Mehra, S. and Nagabhushanam, V. (1973). Complete duplication of bladder and urethra. *J. Urol.*, **109**, 512

36 Redman, J. F. and Bissada, N. K. (1975). Complete duplication of the urethra with probable diphallus. *J. Pediatr. Surg.*, **10**, 135

# 9
# Uterine anomalies: clinical significance

R. E. HARRIS

## INTRODUCTION

Significant clinical implications are associated with uterine anomalies. This subject has stimulated interest, debate and controversy over the past 300 years, especially as to the clinical implications of each anomaly. Depending upon the type of anomaly present, varied clinical presentations of hazardous obstetric and gynaecological complications are encountered. The incidence of uterine anomalies (both major and minor) has been reported from a low of 1 in 4000 deliveries to as high as 12% of different obstetric populations studied. However, as the majority are minor abnormalities, they are usually undetected. The least marked deviation from normal is the uterus arcuatus, which is the most common and the most frequently overlooked. Infertility patients have been found to have a higher than expected incidence (4%) of significantly abnormal uteri. The usual incidence of anomalies is approximately 0.6%, whenever there is a high index of clinical suspicions[1].

Polishuk and Ron[2] reported on three families with two or more sisters having identical Müllerian duct fusion defects. These related patients were detected because all had consulted the same gynaecologist for their previously poor obstetric histories. It may be that the familial (or genetic) nature of uterine anomalies is more frequent than has been realized due to the frequent failure of clinical suspicion and diagnosis. Wiersma et al[3]. reported a mother–daughter relationship with the same types of uterine and urinary tract abnormalities. Therefore, prior knowledge of the existence of uterine anomalies in other members of the family could have important implications in antenatal care and management of related patients, who should be investigated for possible uterine fusion defects whenever there are any clinical indications.

Strassman[4] suggested that congenital abnormalities of the female reproductive system are among the most frequently overlooked causes of obstetric and gynaecological problems. Because the clinical presentations of an anomaly may occur at any time in life by diverse ways, the diagnosis is often not considered without a strong index of suspicion by the clinician. If an anomaly is not considered, it is not sought and would be only fortuitously found.

There is no good explanation as to why some double uteri cause reproductive loss while others provide a series of normal pregnancies. Thus, identification of those double uteri which cause problems can only be made after the fact by clinical trial.

Reproductive failure may be due to a variety of factors. These factors must be eliminated as the cause for the individual patient prior to the consideration of surgical corrections of any uterine defects. Figure 9.1 is a radiograph of a hysterosalpingogram which demonstrates this point.

## EMBRYOLOGY

Uterine anomalies result when there is a complete or incomplete failure of fusion of the Müllerian ducts. The fused portion of the Müllerian ducts gives rise to the uterus and four fifths of the vagina while the unfused portion gives rise to the fallopian tubes. As the most significant events in the embryological formation of the reproductive tract commence before the 6th week of fetal life, the basic causes for a partial or complete failure of Müllerian duct fusion occur during this period. Both fetal and environmental factors of maternal or placental origin have been implicated as the underlying cause but no one factor has been universally accepted. Poor nutrition, metabolic disorders, viral diseases, placental anomalies and exposure to toxic drugs are all included in these factors. However, most women with Müllerian duct fusion defects are

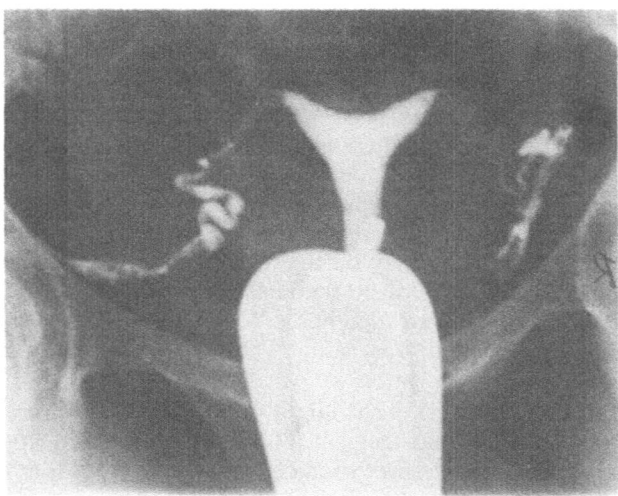

**Figure 9.1**  28-year-old P 1–0–5–1 had five spontaneous abortions. However, the cause was not the arcuate uterus demonstrated here. Without surgical correction, she carried the pregnancy to term

otherwise normal, with the exception of associated urinary tract anomalies such as renal agenesis, horse-shoe kidney and ectopic kidney (Figures 9.2 and 9.3).

Wiersma *et al*[3]. emphasized the importance of investigating the urinary tract in the presence of uterine anomalies. Frequently, these two developmental system anomalies coexist. Two thirds of the patients studied subsequently developed serious kidney disease. Most often uterus bicollis bicornis (with a partial vaginal septum) is associated with unilateral renal agenesis. In their review of the literature, an obstructive uterine anomaly was almost always associated with an absent kidney. For this combination of urological and uterine anomalies to occur, the teratogenic action must be active from the 4th to the 11th week of embryonic life.

Farber and Mitchell[5] reported partial atresia of the fallopian tubes discovered in two patients with double uteri of the bicornuate variety. They pointed out the potential for the coexistence of congenital malformations of fallopian tubes with various uterine malformations, suggesting that extreme caution be exercised during exploratory surgery in these patients with anomalous development of Müllerian ducts.

As to embryological timing, complete failure in development of the uterus

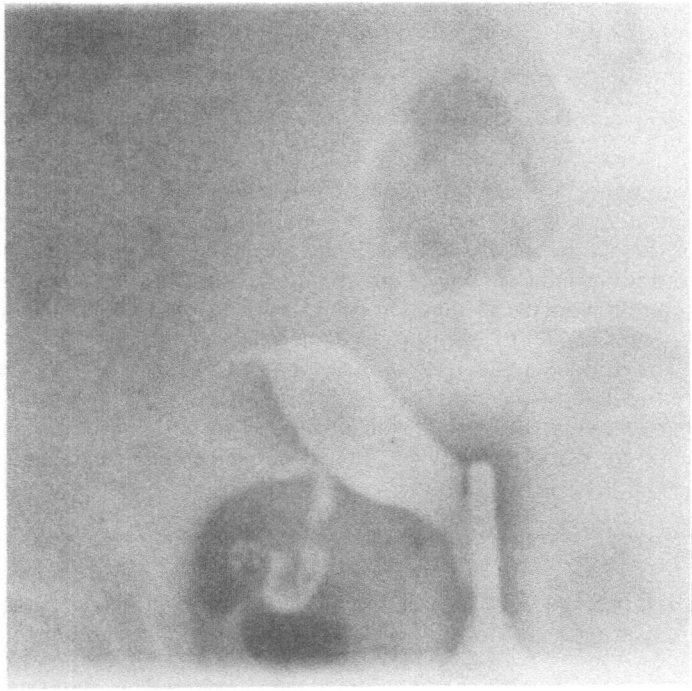

**Figure 9.2** 26-year-old P 0–2–0–2 delivered both infants prematurely at 36 weeks gestation. Both were breech presentations. Postpartum hysterosalpingogram revealed a unicollis unicornis

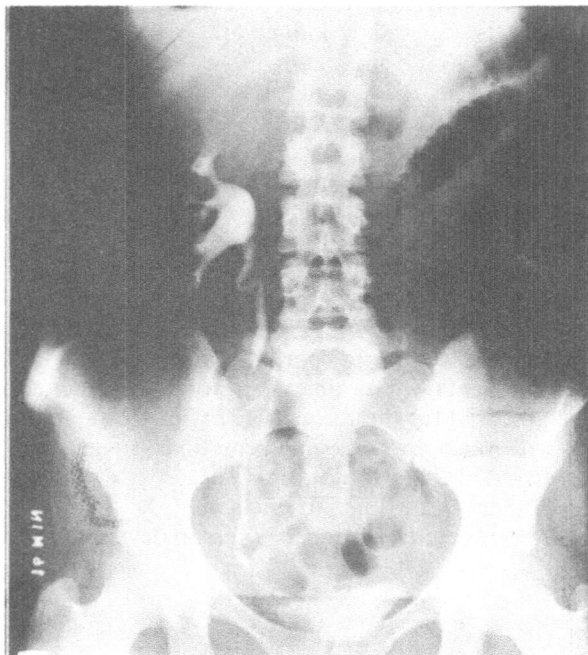

**Figure 9.3** Intravenous pyelogram of the patient in Figure 9.2, showing absent left kidney

occurs within the first embryonic month, whereas complete duplication occurs due to causative factors prior to 10 weeks of life. These factors result in faulty fusion of the Müllerian ducts. Less severe anomalies usually develop during the 3rd month of fetal life due to incomplete fusion. By 10–11 weeks of fetal life, fusion of the ducts has begun to occur to form a single genital canal. Yet, up to 16 weeks of life, delayed or incomplete fusion will result in an arcuate uterus. The fundal septal defect has poor vascularity which has been implicated as one of the causes of fetal wastage since the septum does not provide an ideal site for implanation. Placental growth and development, as well as fetal nutrition and oxygen exchange, are inhibited by this deficit.

## CLASSIFICATIONS OF UTERINE ANOMALIES

The earliest classifications of anomalies were on the basis of anatomical findings and embryological development. These categories were cumbersome with varied groups and were of little benefit to the clinician who needed a simple measurement to evaluate the occasional patient who presented with a perplexing problem. Efforts to compare results of patient evaluations have often been fragmented or present conflicting observations due to varying classification categories.

Jarcho's classification[6] was an early approach to the clinical classification

(Table 9.1). Later, Jones[7] presented a functional classification based upon uterine, cervical and vaginal capacity. In 1962, Semmens[8] proposed a simple classification based upon uterine anatomical and physiological characteristics (Figure 9.4). This classification presents variations of uterine architecture with potential space for the developing fetus combined with those physiological abnormalities which hinder normal growth and development of the fetus, and/or contribute to premature or desultory labour, or abnormal fetal presentation.

Group I consists of functional uteri of single Müllerian origin with a smaller capacity – which includes uterus bicornis bicollis, uterus unicornis and uterus bicornis unicollis with rudimentary horn. Group II consists of functional uteri of dual Müllerian origin with varying degrees of failure of fusion or absorption of the medial septa – which includes uterus bicornis unicollis, uterus subseptus and uterus arcuatus. In 1979, Buttram and Gibbons[9] proposed a new classification, modified after Jarcho's original presentation,

Table 9.1  Types of uterine anomalies according to Jarcho

1 Uterus didelphus
2 Uterus duplex bicornis bicollis
3 Uterus bicornis unicollis
4 Uterus septate
5 Uterus subseptate
6 Uterus arcuatus
7 Uterus unicornis
8 Uterus bicornis unicollis (with rudimentary horn)

UTERUS
BICORNIS BICOLLIS

UTERUS UNICORNIS

UTERUS
BICORNIS UNICOLLIS
ONE HORN RUDIMENTARY

GROUP I. FUNCTIONAL UTERI OF SINGLE MÜLLERIAN ORIGIN

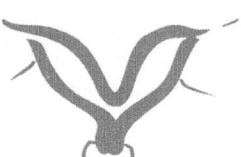 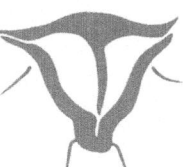 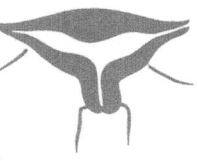

UTERUS BICORNIS UNICOLLIS
BICORNUATE UTERUS          UTERUS SUBSEPTUS   UTERUS ARCUATUS

GROUP II. FUNCTIONAL UTERI OF DUAL MÜLLERIAN ORIGIN WITH VARYING
DEGREES OF FAILURE OF FUSION OR ABSORPTION OF THE MEDIAL
SEPTA

Figure 9.4  Functional classification of uterine anomalies

devised to offer clarity and uniformity based upon the degree of failure of normal development (Figure 9.5). This separates the anomalies into the groups with similar clinical manifestations, treatment and prognosis for fetal salvage. Class VI of their classification has been designated for uteri with internal luminal changes, such as demonstrated secondary to *in utero* exposure to diethylstilbestrol (DES). None of the previous classifications have

### CLASS I: "MÜLLERIAN" AGENESIS OR HYPOPLASIA

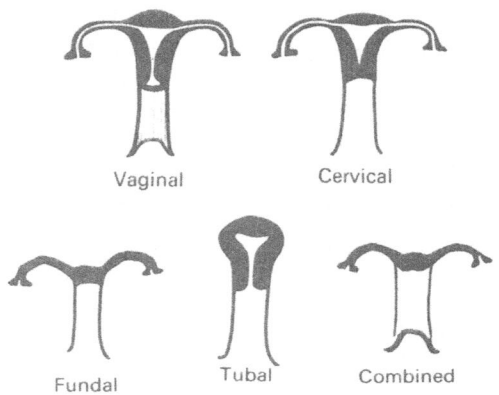

Vaginal        Cervical

Fundal    Tubal    Combined

### CLASS II: UNICORNUATE UTERUS

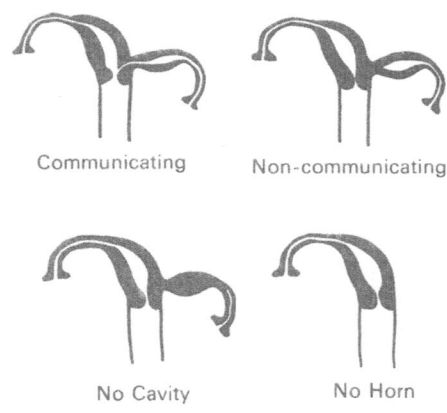

Communicating    Non-communicating

No Cavity      No Horn

### CLASS III: UTERUS DIDELPHYS

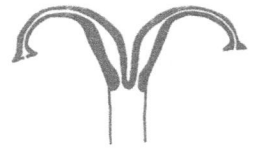

CLASS IV: UTERUS BICORNUATE

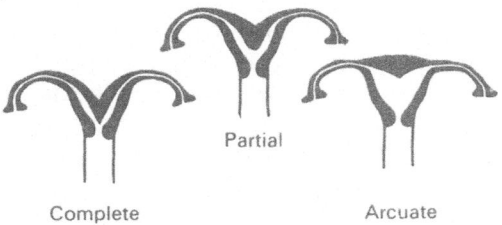

Partial

Complete                    Arcuate

CLASS V: SEPTATE UTERUS

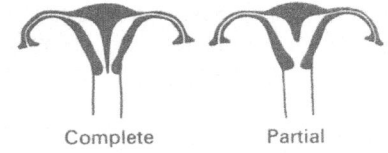

Complete          Partial

CLASS VI: DES ANOMALIES

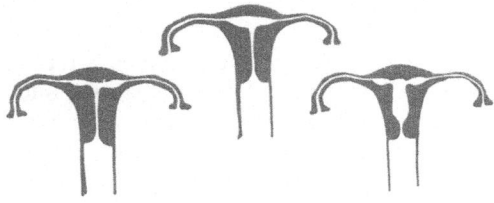

**Figure 9.5** Classification of uterine anomalies proposed by Buttram and Gibbons[9]

so noted this group of anomalies. These typical changes secondary to DES exposure differ from the classic Müllerian malformations.

Whenever a patient has an anomalous uterus, the clinician should specify and describe the type of anomaly, not just by a group, in order to avoid misunderstandings. In addition, when classifying patients, the system being used should be specified. As more data become available, these classification systems should become more widely standardized and recognized by clinicians.

## CLINICAL INDICATIONS TO SUSPECT THE DIAGNOSIS OF UTERINE ANOMALIES

The diagnosis of uterine anomalies is only made when there is an awareness and suspicion of the potential coexistence with specific clinical problems.

Green and Harris[1] particularly emphasized this by demonstrating differences in reported incidences as a reflection of the degree of clinical suspicions and personal interest of the clinicians rendering primary health care (Figure 9.6).

A systematic approach to patient care will make possible the detection of most anomalies. The initial antenatal examination may reveal the first evidence of a potential problem. For example, the patient who has two cervices or two vaginas is very likely to have a double uterus. The status of the cervix is that of an aid to the diagnosis rather than a deterrent to delivery, i.e. whether it is of single or double Müllerian origin. The presence of an asymmetrical located cervix in the vaginal fornix, an excessively large cervix, with or without a septum, or a duplicated cervix in the nullipara warrants investigation. Although it is important to be aware of the possible existence of an anomaly, knowing the exact nature of the anomaly can lead to anticipation of most related complications of pregnancy. Abnormal uterine bleeding is the most common problem which occurs with an anomalous uterus and the incidence of threatened abortion and overall fetal wastage is high.

An increased interest in the diagnosis of these anomalies is essential for understanding and treatment of the clinical problems which occur. Semmens[10] emphasized the correlation of congenital anomalies of the uterus with abnormal fetal presentations detected simply by palpation of the abdominal contour of the pregnant uterus (Figure 9.7 and Table 9.2). As the fetus is frequently in an abnormal position secondary to an abnormal contour of the uterus, the clinician should be suspicious of the presence of a congenital anomaly of the uterus.

At the time of labour and delivery, several clinical situations are associated with uterine anomalies. Whenever premature labour and delivery occurs, postpartum manual exploration of the uterus should be done to detect anomalies. Prolonged labour, as with a breech presentation, may result in

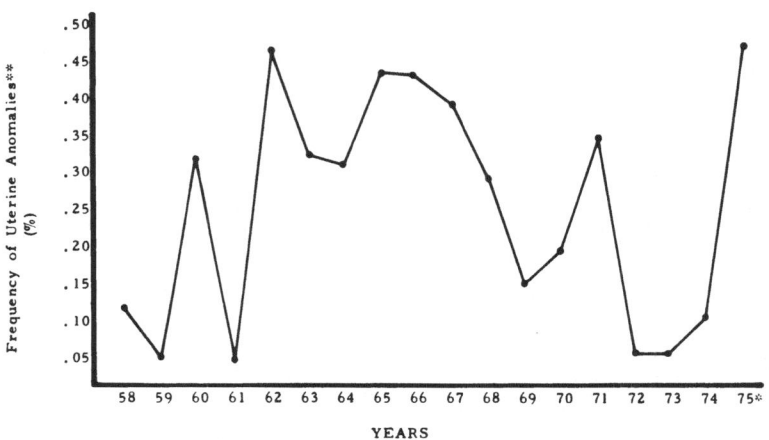

*January to July 1975
**The total number of deliveries during the time period was 31,836

**Figure 9.6** Frequency of uterine anomalies diagnosed per year. (*From* Green and Harris[1])

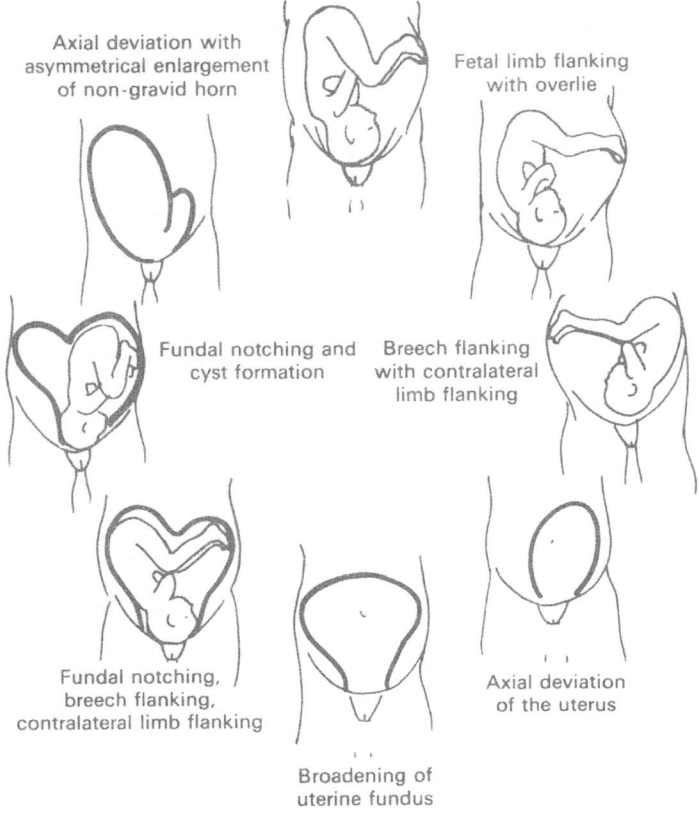

Figure 9.7  Diagnostic signs of uterine anomalies. (*From* Semmens[10])

**Table 9.2  Clinical findings on antepartum abdominal examination suggesting uterine anomalies[10]**

1 Fetal limb flanking with overlie
2 Breech flanking with contralateral limb flanking
3 Fetal limb flanking with fundal notching
4 Broad fundus with fundal notching
5 Cystic formation of uterus
6 Persistent breech presentation
7 Axial deviation of the uterus
8 Floating vertex at term in primigravida

perforation of the septa – preventing descent of the presenting part. The incidence of placenta previae appears to be increased. Considerable postpartum bleeding, which may occur, is secondary to the septal area of the uterus, cervix or vagina. In addition, retained, or even trapped, placenta may occur. With manual removal of the placenta, triangular spasm and cornual pocketing develops, which is increased by administration of oxytocics.

Postpartum, the diagnosis can be made by hysterogram or hysterosalpin-gogram or, perhaps at the time of dilatation and curettage for retained fragments, a septum may be detected. Other clinical indications which indicate the need for diagnostic evaluation include habitual aborters and those patients with renal defects, e.g. absent, fused or ectopic kidney.

Young and Gibson[11] emphasized the importance of correlating hys-terographic studies with the findings of laporoscopic and/or laparotomy visualization. Their patients, evaluated for fertility problems, were noted on hysterograms to have unicornis uteri but to have on laparotomy normal symmetrical external uterine surfaces and bilateral fallopian tubes. They suggested that these patients represented a subgroup of patients with a unicorni cavity not included in Buttram and Gibbons' proposed classification[9] of Müllerian anomalies. Others have indicated that the external contour has little relation to the functional capacity of the uterus.

Nickerson[12] undertook an evaluation to determine if there were a correlation between an abnormal hysterographic contour in primary infer-tility and subtler anomalies rather than severe fusion defects. He de-monstrated a high incidence of uterine anomalies for those patients with primary infertility when no other aetiology had been found.

## CLINICAL COMPLICATIONS OF UTERINE ANOMALIES

The complications of abnormal uteri are variable throughout life (Table 9.3). It is reported that of these patients, 24–53% will abort, 15% will have premature deliveries, and nearly half present with premature rupture of the membranes. In addition, 1 of 6 patients will have abnormal fetal presentations and 1 of 5 will have uterine inertia. Intrapartum complications include septal dystocia, uterine inertia, prolapsed umbilical cords secondary to abnormal fetal presentations, and incarceration of the non-gravid horn. In the management of these patients who present with fetal malpresentations, delivery by caesarean section should be liberally utilized. For those who have had previous corrective operative metroplasty procedures to repair the septal defects (e.g. Strassman), delivery by caesarean section has been the preferred route of delivery[13].

Musich and Behrman[14] were able to show an improvement from a 7% successful pregnancy rate preoperatively to 75% postoperatively improve-ment in obstetric survival. Strassman, Jones or Tompkins procedures were performed according to the type of anatomical defect that was detected. A

Table 9.3  Variable clinical presentations reported with uterine anomalies

| 1 | At birth | Hydrometra, lower vaginal and urinary tract anomalies |
|---|---|---|
| 2 | At puberty | Amenorrhoea, dysmenorrhoea, menorrhagia |
| 3 | At coitus | Dyspareunia, intrauterine device failures |
| 4 | Infertility | Primary, secondary |
| 5 | During pregnancy | Abortion, ectopics, premature labour, abnormal fetal presentation, haemorrhage |
| 6 | During labour and post partum | Dystocia, retained placenta, postpartum haemorrhage |
| 7 | Later life | Endometriosis, menorrhagia |

causal relationship between uterine deformity and pain, bleeding and lack of conception is supported only by anecdotal evidence. Only reproductive failure, as evidenced by a specific kind of spontaneous abortion, is considered an indication for surgical repair.

In general, if the patient has a single cervix, the antepartum course is complicated but she should have an uncomplicated labour. However, if the patient has a double cervix, the antepartum course is usually benign, but the labour is complicated. For those patients with a double uterus and vaginal septum, it is best to incise the septum while non-pregnant. However, early in the pregnancy, or preferably at delivery, septal incision can be accomplished. It is better to visually incise the septum in order to control the bleeding, rather than having the fetal presenting part tear the septum and cause extensive bleeding which may be difficult to control. A patient may have two vaginas, one of which terminates in a blind pouch. In this instance, the blind pouch vagina is an effective contraceptive and the couple may not desire the removal of the septum. For this particular situation the septum would not cause any complications – as it would be compressed laterally at the time of delivery – and does not need to be incised.

Labour for the patient with an anomalous uterus should be allowed to progress normally unless inertia develops or an abnormal fetal presentation exists which requires caesarean section delivery. The fetal and perinatal mortality associated with breech presentations and abnormal uteri is very high. Thus these patients may be best managed by caesarean section, especially if double cervices are also present. Patients who develop uterine inertia should not have oxytocic augmentation of labour but rather should be delivered by caesarean section. If the patient has had a previous fetal loss during labour, delivery by caesarean section should be considered and discussed with her prior to the time she is admitted in labour with the present gestation.

In some instances, those patients whose anomalies are in functional group I category have precipitous labours and deliveries.

Postpartum complications include uterine atony, adherent placentas and failure of the uteri to involute. Almost one third of patients have retained placentas and postpartum bleeding occurs commonly. Manual exploration of the uterine cavity is recommended to attempt to detect anomalies whenever the placenta does not separate promptly or for patients who deliver prematurely. Wiebe[15] reported an instance of retained placenta in the right cornu of a bicornuate uterus. The developing fetus had extruded into the left cornual area and developed to maturity. However, the placenta was trapped behind the non-yielding opening of the right cornual area and had to be removed at laparotomy.

Rudimentary uterine horn has been reported to be associated with intrauterine fetal deaths. Patients with such an anomaly may present as an early ectopic pregnancy or the pregnancy may progress until the uterine horn ruptures with massive internal haemorrhage and death. Rupture of the rudimentary horn has been reported as early as the 10th week or as late as the 38th week of gestation. The average duration of pregnancy for a rudimentary horn prior to rupture has been $21\frac{1}{2}$ weeks. Cohn and Goldenburg[16] reported a maternal survival following an instance with catastrophic effect, secondary to fetal implantation in a rudimentary uterine horn. Rarely do pregnancies in rudimentary horns yield viable offspring[17].

One interesting feature of this type of anomaly is that the patient may not labour, despite the fact that in one reported instance the patient was over 44 weeks gestation and delivered a postmature infant. With the removal of the rudimentary horn, every effort should be made not to remove the remaining functional uterus and normal adnexae.

Johansen[18] reported near-fatal courses for patients with ruptured cornual pregnancies in rudimentary horns which dramatized the dangers of such implantations during pregnancy.

Some women with rudimentary uterine horns will give a history of previously normal pregnancies (Figure 9.8). In addition, a number had become normally fertile after the removal of the rudimentary horn. Amputation of the rudimentary horn which contains a pregnancy should and can be

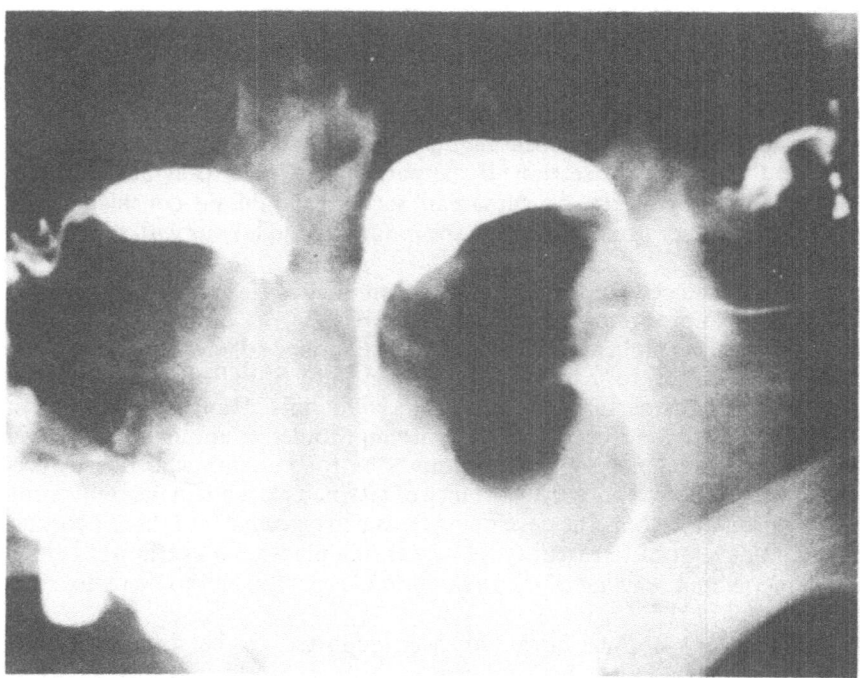

**Figure 9.8**  19-year-old P 0101 delivered at 37 weeks and noted antenatally to have an abnormal abdominal contour. Postpartum hysterosalpingogram reveals a communicating rudimentary horn

done safely. If a patient is operated upon for another reason and a rudimentary uterine horn is found, it should be removed at the time of the exploratory surgery. Johansen recommended elective removal of such a vestigial horn whenever it is identified prior to the gestation. This seems to be reasonable prophylaxis to prevent this catastrophic problem.

Szlachter and Weiss[19] reported segmental tube absence in conjunction with Müllerian duplication in which the patient had a ruptured ectopic pregnancy in the blind distal segment of the right tube. They compared this, as being similar, to a non-communicating rudimentary horn. Thus, if a non-communicating tube were present, with Müllerian reduplication, the tube should be either reanastomosed or removed to prevent ectopic pregnancy complications.

A combination of twin pregnancy and abnormal uteri can present unexpected problems. Pelosi et al.[20] reported a patient who desired termination of her pregnancy. This procedure was performed and was considered successful. However, it was not detected that the patient had a double uterus. As the pregnancy continued in the other uterine cavity, she had to return for an elective therapeutic abortion of the other cavity. McArdle[21] reported a patient who had two attempts at therapeutic abortion with failure due to the curettage being performed on the non-gravid cornu of the septate uterus. Loendersloot[22] reported failure of suction curettage when the pregnancy occurred in a rudimentary horn. Thus, both sides of the uterus should be evaluated at the time of therapeutic abortion.

Other gynaecological complications have been reported, e.g. dyspareunia, dysmenorrhoea, menorrhagia, haematocoele and hidden carcinoma in one horn of the uterus. Patients with a rudimentary horn have been reported to have severe and disabling dysmenorrhoea two to three times that of the normal population. Thus, patients with intractable abdominal pain, especially during menses, should be considered as possibly having an uterine anomaly[23].

Neves-E-Castro et al.[24] reported obstruction of one side of the vagina with a lateral communication between the two halves of a double uterus above the level of the external cervical os. This created a special situation, clinically puzzling but threatening future reproduction if not recognized and corrected. There are several distinct features for this anomaly. It is not easily recognized, instrumentation for diagnostic purposes is apt to induce infection, and renal agenesis is noted on the side of the obstructed hemivagina. The vaginal cystic mass is a haematocolpos of the right occluded vagina.

Following in utero DES exposure, there have been increasingly reported associations with a high percentage of abnormal hysterograms.

Of Kaufman and co-workers'[25] 60 patients with known DES exposure and abnormal hysterosalpingograms, the most common alteration was a T-shaped appearance of the uterine cavity with a widening of the isthmic portions of the oviduct. In other reports, there were constriction bands around various portions of the uterine cavity. For some patients, a widening of the lower portion of the endometrial cavity was observed, as well as occasionally a combination of all effects[26,27].

175

Thus, during the initial evaluation of pregnant patients, an inquiry concerning possible intrauterine DES exposure is important. Patients with such a history should have closer surveillance for complications during the gestation. Although relationship with DES has not definitely been proved, complications such as ectopic pregnancies, insufficient cervix, abnormal fetal presentations, placenta previaes, premature labours, retained placentas, and postpartum haemorrhages have been reported to occur more frequently in patients who had *in utero* DES exposure. As previously noted, all of these complications are also associated with those patients who have abnormal uteri. Once again, anticipation and recognition of these interrelationships could lead to earlier detection, possible correction of defects and improvement in fetal and perinatal outcomes.

Behnam[28] reported that there have been so few cases of vaginal carcinoma associated with congenital malformations of the uterus that any considered relationship would be coincidental.

However, it is important that a patient with a double cervix have a pap smear obtained from each cervical os. It should be recognized that such a patient can have cervical carcinoma in one cervix[29]. In other words, the pap smear obtained from one cervix may be normal while the carcinoma is continuing unnoticed in the other cervix. In addition, carcinoma may be present in both cervices[30].

The ratio of endometrial carcinoma to cervical carcinoma in anomalous uteri is higher than would be expected[31]. Delay in diagnosis and treatment of endometrial carcinoma occurring in one horn of a double uterus has been reported. Thus, whenever evaluating a patient for abnormal uterine bleeding, it is important to know the presence of a double uterus in order to perform adequate diagnostic procedures.

## CONCLUSION

Patients with uterine anomalies generally seek help because of repeated gestational losses and infrequently for such symptoms as pain or bleeding. Infertility is rarely the presenting complaint, nor should it be an operative indication. Patients who have uterus didelphis, more often than not, have satisfactory obstetric histories and are unlikely to require repair. Contrary to the opinion of some, the didelphic condition seems to offer the best prognosis for successful pregnancy of all the anomalies. Patients with septate bicornuate uteri are poor obstetric performers but if diagnosed, properly selected, and surgically repaired, can be expected to achieve excellent obstetric results.

A high index of suspicion that an anomalous uterus is the underlying cause for a clinical problem is essential for accurate diagnosis and therapy.

## References

1 Green, L. K. and Harris, R. E. (1976). Uterine anomalies, frequency of diagnosis and associated obstetrics complications. *Obstet. Gynecol.*, **47**, 427

2  Polishuk, W. Z. and Ron, M. A. (1974). Familial bicornuate and double uterus. *Am. J. Obstet. Gynecol.*, **119**, 982

3  Wiersma, A. F., Peterson, L. F., and Justema E. J. (1976). Uterine anomalies associated with unilateral renal agenesis. *Obstet. Gynecol.*, **47**, 654

4  Strassman, E. O. (1966). Fertility and unification of double uterus. *Fertil. Steril.*, **17**, 165

5  Farber, M. and Mitchell, G. W. Jr. (1979). Bicornuate uterus and partial atresia of the fallopian tube. *Am. J. Obstet. Gynecol.*, **134**, 881

6  Jarcho, J. (1946). Malformations of the uterus. *Am. J. Surg.*, **71**, 106

7  Jones, W. S. (1957). Obstetric significance of female genital anomalies. *Obstet. Gynecol.*, **10**, 113

8  Semmens, J. P. (1962). Congenital anomalies of female genital tract: Functional classification based on review of 56 personal cases and 500 reported cases. *Obstet. Gynecol.*, **19**, 328

9  Buttram, V. C. Jr. and Gibbons, W. E. (1979). Müllerian anomalies: a proposed classification (an analysis of 144 cases). *Fertil. Steril.*, **32**, 40

10  Semmens, J. P. (1965). Abdominal contour in the third trimester – an aid to diagnosis of uterine anomalies. *Obstet. Gynecol.*, **25**, 779

11  Young, O. H. and Gibson, M. (1980). Unicornis uterus with a normal external uterine surface. *Fertil. Steril.*, **33**, 663

12  Nickerson, C. W. (1977). Infertility and uterine contour. *Am. J. Obstet. Gynecol.*, **129**, 268

13  Buttram, V. C. Jr., Zanotti, L., Acosta, A. A., Vanderheyden, J. S., Besch, P. K. and Franklin, R. R. (1974). Surgical correction of the septate uterus. *Fertil. Steril.*, **25**, 373

14  Musich, J. R. and Behrman, S. J. (1978). Obstetric outcome before and after metroplasty in women with uterine anomalies. *Obstet. Gynecol.*, **52**, 63

15  Wiebe, D. (1970). Retained placenta of unusual type. *Obstet. Gynecol.*, **35**, 153

16  Cohn, F. L. and Goldenburg, R. L. (1976). Term pregnancy in an unattached rudimentary uterine horn. *Obstet. Gynecol.*, **48**, 234

17  Jarrell, J., Effer, S. B. and Mohide, P. T. (1977). Pregnancy in rudimentary horn with fetal salvage. *Am. J. Obstet. Gynecol.*, **127**, 676

18  Johansen, K. (1969). Pregnancy in a rudimentary horn. Two case reports. *Obstet. Gynecol.*, **34**, 805

19  Szlachter, N. and Weiss, G. (1979). Distal tubal pregnancy in a patient with a bicornuate uterus and segmental absence of the fallopian tube. *Fertil. Steril.*, **32**, 602

20  Pelosi, M. A., Langer, A., Li, T. S., Zanvettor, J. and Cortes, R. (1977). Failed termination of pregnancy due to uterus bicornis unicollis with bilateral pregnancy. *Am. J. Obstet. Gynecol.*, **128**, 919

21  McArdle, C. R. (1978). Failed abortion in a septate uterus. *Am. J. Obstet. Gynecol.*, **131**, 910

22  Loendersloot, E. W. (1977). Twin pregnancy in double uterus. *Am. J. Obstet. Gynecol.*, **127**, 682

23  Rolen, A. C., Choquette, A. J. and Semmens, J. P. (1966). Rudimentary uterine horn. Obstetric and gynecologic implications. *Obstet. Gynecol.*, **27**, 806

24  Neves-E-Castro, M., Saavedra, A. B. E., Vilhena, M. M. and Jones, H. W. Jr. (1976). Lateral communicating double uterus with unilateral vaginal obstruction. *Am. J. Obstet. Gynecol.*, **125**, 865

25  Kaufman, R. H., Binder, G. L., Gray, P. M. and Adam, E. (1977). Upper genital tract changes associated with exposure in utero to diethylstilbestrol. *Am. J. Obstet. Gynecol.*, **128**, 51

26  Haney, A. F., Hammond, C. B., Soules, M. R. and Creasman, W. T. (1979). DES induced upper genital tract abnormalities. *Fertil. Steril.*, **31**, 142

27  Pillsbury, S. G. Jr. (1980). Reproductive significance of changes in the endometrial cavity associated with exposure *in utero* to diethylstilbestrol. *Am. J. Obstet. Gynecol.*, **137**, 178

28  Behnam, K. (1980). Congenital malformations and carcinoma. *Female Patient*, **5**, 42

29  Gerbie, M. V. and Weingold, A. B. (1966). Cervical carcinoma in an anomalous cervix. *Obstet, Gynecol.*, **27**, 168

30  Wall, R. L. Jr. (1958). Didelphic uterus with carcinoma in situ both cervices. *Am. J. Obstet. Gynecol.*, **76**, 803

31  Duncan, A. S. and John, A. H. (1962). Endometrial carcinoma occurring in the uterus bicornis. *J. Obstet, Gynecol. Br. Commonw.*, **69**, 488

# 10
# Congenital anomalies of the vagina

C. A. SALVATORE

Congenital anomalies of the vagina are represented by vaginal agenesis or atresia associated with uterine agenesis (Rokitansky–Kuster–Hauser Syndrome), vaginal agenesis with functioning uterus, septate vagina, double vagina and wolffian cystic formation (Gartner's cyst). Agenesis is the term used for complete absence of the vagina and atresia when a fibrous cord exists at the site of the vagina.

The most common anomalies are represented by vaginal agenesis which influences sexual relations. Congenital agenesis exerts a greater influence on a woman's personality, sexuality and acceptance by her partner. It is a legal, social and medical problem requiring treatment at the proper time in a woman's sexual evolution.

The vagina is the end portion of the internal female genital organs, the birth canal, and is the passage for the flow of cervical and menstrual secretion.

## DEVELOPMENT OF THE VAGINA

It is essential to review here the embryology and morphology of the vagina. In a 42 mm female fetus, when the posterior ends of the Müllerian ducts reach the urogenital sinus, there is contact with the wolffian ducts. The two Müllerian ducts also fuse into a uterovaginal canal. This is prolonged by a solid cord of clear cells, the 'vaginal cord', up to the vestibule. As the wolffian ducts retrogress, the vaginal cord shows a central lumen. Several studies[1,2] have shown that the participation of the Müllerian ducts in the vaginal cord occurs only in the upper part of the vagina. At least the major part of the vagina is derived from the urogenital sinus[3].

The epithelium of the urogenital sinus progressively extends around the posterior part of the wolffian ducts which enlarge and will constitute the vaginal plate. The outgrowths from the sinus are double in origin – one around each wolffian duct – but posteriorly they fuse under the Müllerian ends. It seems that the urogenital sinus is induced to proliferate within the lumen and outside of it at the area of contact with the Müllerian ducts.

The major part of the vaginal cord first increases in diameter and acquires a lumen near the urogenital sinus. The fornices and the lumen of the whole genital tract appear at the 200 mm stage of the fetus.

The development of the uterus precedes that of the vagina. For this reason, most cases of vaginal agenesis are usually found in association with uterine agenesis. However, the vagina can be partially formed in the absence of the uterus and vice versa, although this is rare. Other congenital anomalies, especially of the urinary tract, are commonly found in association with those of the genital system.

After birth, the vagina together with the uterus continues developing slowly. Later, during adolescence, this development is more rapid.

Normally, the vagina is found between the urethra–bladder and the rectum posteriorly. The vagina is attached to the lateral pelvic wall by connective tissue condensation and smooth muscle adherent to the adventitia of the vaginal blood vessels. There is a large amount of elastic tissue which permits great distensibility with subsequent return to the previous state, as is observed during and after labour.

Radiographic colpography done by Nichols and Randall[4] shows a distinct superiorly convex perineal curve in the lower vagina. The upper vagina is parallel to the levator ani muscles. The rectum, vagina and urethra pass through the hiatus of the pelvic diaphragm. Under normal conditions the cardinal and uterosacral ligaments are important for maintaining not only the uterus but also the upper vagina.

## PATHOGENESIS

The aetiology of the congenital anomalies of the vagina is not clearly understood. In general, there are no chromosomal anomalies in relation to vaginal agenesis (Rokitansky's syndrome)[5]. However, three cases of vaginal agenesis have been reported in three monozygotic twins with normal twin sisters. They have been reported in relation to genetic aetiology by Linscke et al.[6]. Le Duc et al.[7] described 2 only of 14 cases of vaginal anomaly without karyotype. In the Rokitansky–Kuster–Hauser syndrome, normal 46,XX karyotypes occur. It seems that the congenital malformations depend on environmental factors in the presence of changes in DNA. The DNA mutation can be spontaneous or induced by ionizing radiations, viral or chemical agents which affect the DNA code.

Pathogenesis of vaginal anomalies is related to the development of the Müllerian ducts during the fourth week of embryonic development. During the fifth month, the definitive lumen of the vagina is partially formed. A process of excavation and cellular breakdown begins caudally, giving rise to partial or total vaginal and uterine agenesis. According to the studies of Jost[8,9], Müller's ducts begin to regress or develop when the primitive gonad differentiates into either the fetal testicle or ovary, respectively. On the differentiation of the primitive gonad into an ovary, the wolffian ducts, in the absence of stimulation, will regress, but sometimes persist as small ducts which later give rise to the wolffian vaginal cysts.

Cessation of the development of the entire uterovaginal canal causes vaginal and uterine agenesis (Rokitansky–Kuster–Hauser syndrome)[10]. But, when development of the uterovaginal canal continues beyond the fourth month it causes congenital absence of the vagina with a normal uterus. Intermediary forms of abnormal development of Müller's ducts are much more common, with the rudimentary uterus represented by hemiuteri or just one nodule. Sometimes, the vagina is partially or completely formed but with partial (septate vagina) or total septum. There is also the possibility of Müller's ducts approaching one another but not fusing. This would result in two hemiuteri and a double vagina.

The vaginal vestibule and hymen originating from urogenital sinus in general is normal and explains the common presence of a small retrohymenal fossa and normal hymen in the majority of cases of vaginal agenesis. Thus, according to most authors[11-17], the external genital organs have a normal appearance.

Congenital defects can involve either deficient muscle or nerve supply. The association of vaginal anomalies with renal and bone anomalies is also common. Most patients show renal agenesis or ectopic kidneys, double ureters or just one kidney and ureter.

In a previous report, Salvatore et al.[18] found 4 cases with only one kidney and 2 with double pelvis-calyces (17.8%) in 90 cases of vaginal agenesis. Another publication reported that 28.6% of cases of vaginal agenesis were associated with anomalies of the urinary system[19]. This confirms previous observations[20,21].

## CLASSIFICATION

For practical purposes, agenesis and atresia will be considered together. The vaginal anomalies can be classified as follows.

(1) *Vaginal agenesis*
    (a) Total vaginal agenesis
        (i) Without uterus
        (ii) With rudimentary uterus
        (iii) With double rudimentary uterus
        (iv) With hypoplastic uterus
        (v) With normal functional uterus
    (b) Solid and rudimentary vaginal agenesis (atresia) same as item (a)
    (c) Partial proximal vaginal agenesis
    (d) Partial distal agenesis.

(2) *Vaginal septum*
    (a) Vaginal transverse septum (canalization failure)
    (b) Vaginal longitudinal septum
    (c) Vaginal duplication (total longitudinal septum: double vagina)
        (i) With one cervix
        (ii) With two cervices.

(3) *Vaginal wolffian cysts.*

(4) *Associated anomalies*
  (a) *Uterus*
    (i) Absent or rudimentary
    (ii) Functioning (with cryptomenorrhoea)
  (b) *Ovary*
    (i) Normal
    (ii) Hypoplastic
  (c) *Tubal*
    (i) Hypoplastic
    (ii) Rudimentary
    (iii) Normal
  (d) *Urinary tract*
    (i) Two normal kidneys
    (ii) One normal kidney
    (iii) Left kidney absent
    (iv) Right kidney absent
    (v) Two pelvic kidneys
    (vi) Two ureters.

According to Hafez and Evans[22], we may also find the following associated anomalies.

(1) Musculoskeletal
(2) Mental retardation
(3) Hermaphroditism (we have observed three cases)
(4) Hernia
(5) Other, very rare, associated anomalies of no great importance.

## CLINICAL ASPECTS OF VAGINAL ANOMALIES

### Vaginal agenesis

The clinical aspects of vaginal anomalies are important for the correct diagnosis. The patient's age and history and failure to menstruate up to the time of puberty constitute one of the first important symptoms (primary amenorrhoea).

The development of sexual characteristics, sexual antecedents (attempts at intercourse), presence or absence of libido, and severe periodic pain due to obstruction of menstrual flow also constitute important information for the diagnosis of congenital anomalies of the vagina, obtained by a careful case history.

According to Semmens[23], in a study of more than 20 girls ranging in ages from 16 to 39 with congenital absence of the vagina, the treatment was due to the desire for sexual intercourse. In our cases (90), 24.42 % found it impossible to have sexual relations.

The general physical examination is also important. Bio-type, breast development and hair distribution should be considered. Pimenta Filho[24] found 3 cases of hypotrophic breasts of 33 cases examined. The presence of a palpable tumour in the hypogastrium in young patients with primary

amenorrhoea and severe periodic pain in the hypogastrium indicate haematometra. In 100 cases from the Mayo Clinic[12], 4 were found to have normal functioning uteri. McIndoe[25] observed 9 functioning uteri in 63 cases.

The gynaecological examination provides the diagnosis which in the majority of cases is easily made. As we have mentioned, it is seldom connected with the form of intersex associated with sex chromosome abnormalities[26].

Inspection and examination of the pubic region, labial formation, clitoris, vestibular region and presence or absence of the hymen are important. Generally, these are normal. According to Hauser and Girotti[27], the urethral meatus exposed in the caudal position was observed in 56 % of the cases. The presence of a vaginal cavity can be confirmed by an examination through the hymenal orifice. If the orifice of the hymen is small, it can be explored with a hysterometer. When a terminal vaginal cavity or fossette exists, a little speculum or 'virgoscopium' provides good illumination[28,29]. This method allows good visualization of the cervix and exploration of the vaginal cavity, especially when a septate or double vagina or Gartner's cyst is suspected.

Despite the fact that a simple inspection (Figure 10.1) allows the diagnosis

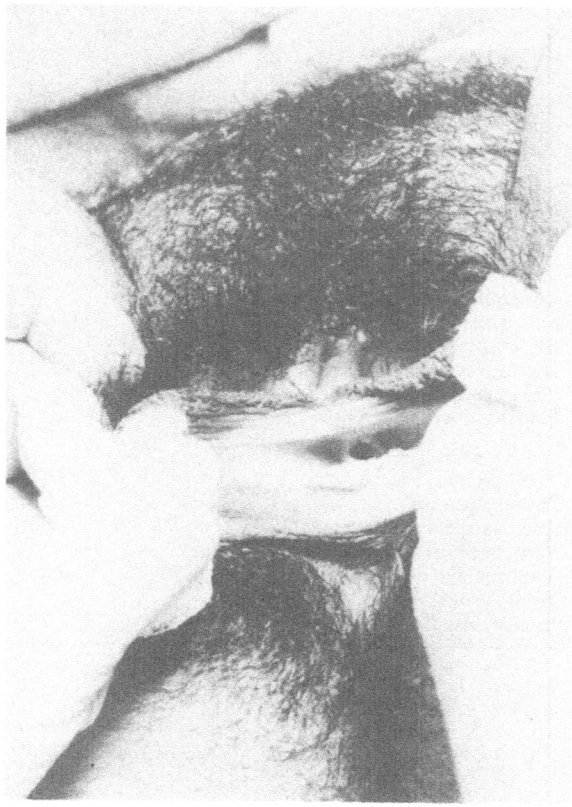

**Figure 10.1**  Vaginal agenesis. Only a ureteral orifice is present

of vaginal agenesis to be made in the majority of cases, a rectal examination is indispensable. It permits identification of the presence or absence of fibrous cords in place of the vagina and the presence or absence of the uterus, which is mainly absent or rudimentary. Even more rarely, it allows one to feel soft and uniform tumoral formations indicating haematocolpos, or haematometra and even haematosalpinx. When the tumour formation is lower, it is suggestive of haematocolpos (partial agenesis) (Figure 10.2), and not haematometra, and vice versa.

On our service there are now more than 110 cases of vaginal agenesis but we will report on 90 cases which were published in collaboration with Lodovici[30]. Details are shown in Table 10.1.

**Table 10.1**

| | No. of cases |
|---|---|
| 1 *Age* (years) | |
| 10–15 | 7 |
| 16–20 | 46 (50%) |
| 21–25 | 24 |
| 26–30 | 10 |
| 31–35 | 4 |
| 2 *Civil status* | |
| Unmarried | 64 (71.1%) |
| Married | 26 (28.9%) |
| 3 *Symptoms* | |
| Amenorrhoea | 90 (100%) |
| Impossibility of coitus | 22 (24.4%) |
| Hypogastric pain | 18 (20.0%) |
| 4 *Gynaecological examination (rectal)* | |
| Non-palpable uterus | 70 (77.7%) |
| Rudimentary palpable uterus | 14 |
| Uterus increased in volume | 4 |
| Unrecognizable ovaries | 80 |
| Palpable ovaries | 10 |
| 5 *Clinical diagnosis* | |
| Total agenesis | 70 (77.7%) |
| Partial agenesis | 10 (11.1%) |
| Total agenesis plus haematometra | 5 (5.5%) |
| Partial agenesis, haematometra and haematocolpos | 3 (3.3%) |
| Total agenesis, haematometra and haematosalpinx | 2 (2.2%) |

## Complementary methods of diagnosis

It is important to explain the cases of congenital anomalies of the vagina, especially in the presence of agenesis.

### Laparotomy (36 cases) and laparoscopy (54 cases)

The exploration of the internal genital organs, uterus, tubes and ovaries is important. Actually, the laparoscopy replaces the laparotomy because it

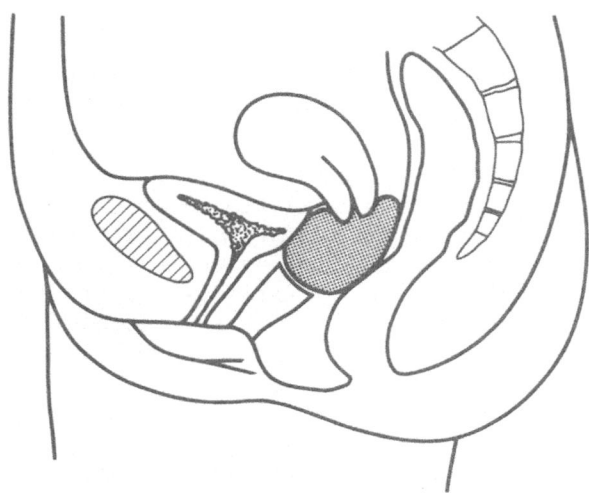

**Figure 10.2**  Haematocolpus in a case of transversal septum of the vagina

enables one to do a biopsy of the ovary and to thoroughly explore the pelvic abdominal cavity. The results found in our material are given in Table 10.2.

**Table 10.2**

|  | No. of cases |
|---|---|
| 1 *Ovaries* | |
| Normal | 68 (75.5%) |
| Polycystic | 6 (6.6%) |
| Hypoplastic | 12 |
| Rudimentary | 4 |
| 2 *Histology of the gonads* (*biopsy 40 cases*) | |
| Normal ovaries | 37 (92.5%) |
| Micropolycystic | 3 |
| 3 *Uterus* | |
| Hypoplastic | 2 |
| Increased in volume (haematometra) | 4 |
| Solid rudimentary | 50 |
| Double uterus | 1 |
| Hemiuterus | 4 |
| Two rudimentary uteri | 4 |
| Lack of uterus | 22 (24.4%) |
| 4 *Tubes* | |
| Normal | 29 (32.2%) |
| Hypoplastic | 6 |
| Rudimentary (long) | 42 (46.6%) |
| One rudimentary tube | 2 |
| Lack of tube | 10 |
| Haematosalpinx | 1 |

The results shown in Table 10.2 are similar to those found by others. Counseller and Davis[20] found 5 functioning uteri in 76 cases (haematometra). Bryan et al.[12], 4 out of 100; Jackson (in Salvatore and Lodovici[30]), 5 out of 128; Cordier[31], 1 out of 23; and Salvatore and Lodovici[30], 7 out of 90 cases (7.7%).

### Pelvigraphy and gynaecography
Pelvigraphy and gynaecography can be used to show the presence or absence of the uterus and ovaries. It was formerly widely used on our service but at present it is reserved for cases of Stein–Leventhal syndrome. Only 6.6% of cases presented polycystic ovaries, contrary to what has been observed by others. For example, Hauser and Girotti[27] have found 60% of polycystic ovaries in cases of vaginal agenesis. We prefer laparoscopy because a biopsy of the ovaries is possible at the same time.

### Excretory urography
Exploration of the urinary system is very important in cases of vaginal anomalies. About 20% of the cases show anomalies of the urinary system. In our series, we found four cases with absence of one kidney, two with double pelvis-calyces and one with double ureters.

### Genetic sex
Determination of the sex chromatin is important, mainly when there are anomalies of the vagina associated with gonadal dysgenesis. In three cases of male pseudohermaphroditism in which artificial vaginas were created, the sex chromatin confirmed a male sex (chromatin absent). Sometimes the Rokitansky–Kuster–Hauser syndrome may be confused with a 'testicular feminizing syndrome'. Nahoun et al.[32,33] found that 7 out of 29 cases of vaginal agenesis had feminizing testicles. As we have stated, the sex chromatin in the Rokitansky–Kuster–Hauser syndrome is always positive and the karyotype is 46,XX.

### Vaginogram
In cases of partial agenesis with stenotic rings, the vaginogram is important for demonstrating the vaginal conformation.

### Sonography
For the differential diagnosis between haematometra and haematocolpus the ultrasound can decide, showing the uterus and the haematocolpus (Figure 10.3).

### Cystoscopy
The cystoscopy is indicated in rare cases of Gartner's cysts to distinguish a urethral diverticulum.

## Vaginal septum and double vagina

Transverse septa can occur at any site of the vagina. The failure of canalization of the vaginal plate or some defect of the vaginal sinus are possible causes of the septa.

Low transversal septa result in haematocolpos. The septa are not complete in some cases and menstruation is still possible.

Longitudinal partial septa are rare. There is a history of painful intercourse and examination with a speculum gives the diagnosis. It shows the presence of

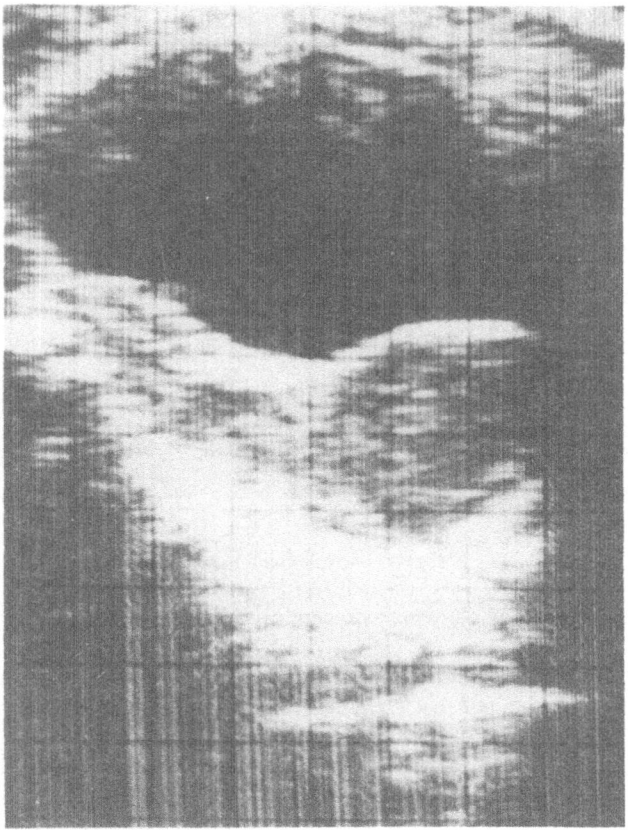

**Figure 10.3** Sonography. Below the bladder, one can see the uterus on the left and the haematocolpus on the right

a single or double longitudinal septum which is often united at the cervix.

A double vagina results when the longitudinal septum is whole and occupies the entire vagina forming two cavities separated by mucosa and muscle layers. This is a rare condition and is commonly associated with a didelphic uterus. Dougherty[34] reported that 20 of 22 patients with didelphic uterus have a double vagina.

A study carried out in the Gynecology Clinic of the University of São Paulo Medical School of 110 cases of vaginal anomalies showed only two cases of double vagina associated with two hemiuteri (Figures 10.4 and 10.5).

## Vaginal cysts

Gartner's vaginal cysts originate from remnants of the wolffian ducts and are found along the lateral vaginal walls. According to Evans and Paine[35], the histology of the cysts includes a muscle layer, and an epithelium with a basal

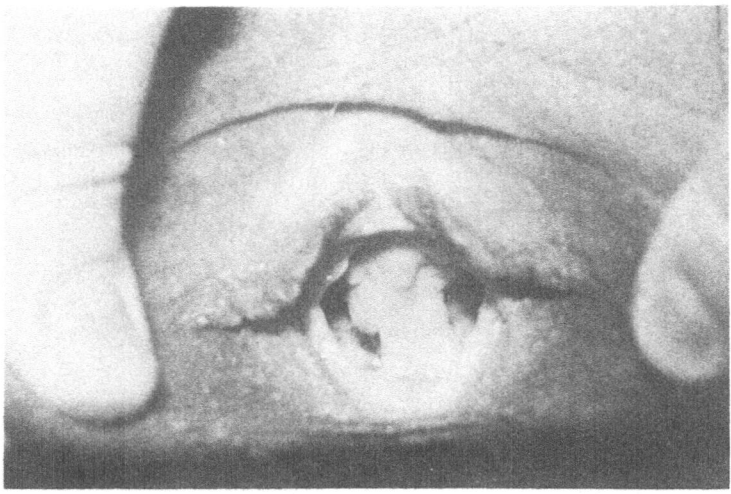

**Figure 10.4** Double vagina

membrane. Sometimes, they may present an epidermoidal metaplasia (Figure 10.6).

Gartner's cysts are usually encountered singly, and on vagino-abdominal examination can be felt on the vaginal wall as a localized spheroid or elongated bulge. Examination with a speculum confirms this (Figure 10.7). They are generally asymptomatic (48.8 % of cases); however, if they are very large they may cause painful coitus (11.8 %).

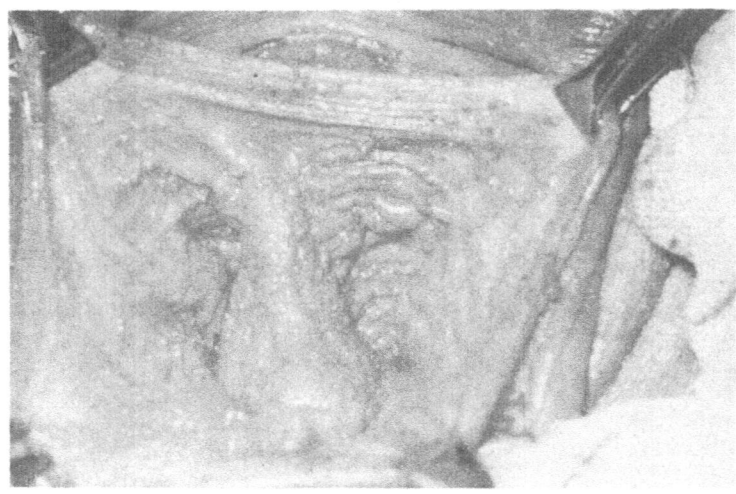

**Figure 10.5** Double vagina

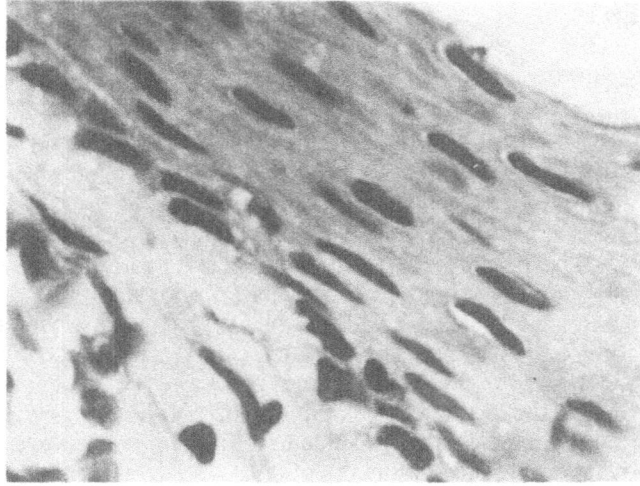

**Figure 10.6** Epidermoidal metaplasia of the columnar epithelium in Gartner's vaginal cysts

On our service the incidence is 0.2%. Of 43 cases of Gartner's cysts seen in our Department and published by Souen and Schivartche[36], none presented malformations originating from Müllerian organs. Similarly, malformations of the urinary system were not noted. The only anomaly of note was the presence of an extra breast in one case. The cyst was located in the right anterolateral wall of the vagina in 44.2% of the cases studied.

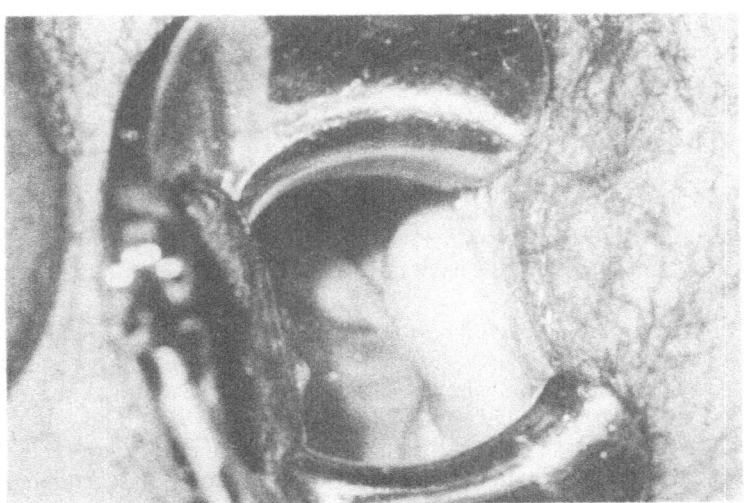

**Figure 10.7** Gartner's vaginal cysts

## TREATMENT

### Vaginal agenesis

Numerous surgical procedures have been devised to treat absence of the vagina. Segment of rectum[37], sigmoid colon[38,39], and terminal ileum[40,41] have been used. These methods in addition to being complex have a high morbidity. Simple pressure may be adequate in very few cases of partial agenesis. The Frank[42-44] technique is not satisfactory. Pedicle grafts have not been useful[44]. The preparation of a cavity in the rectovesical space and waiting for its spontaneous epithelialization offers poor results[45-47]. The best results were obtained by inserting a split-thickness skin graft over a vaginal mould introduced into an appropriate cavity created between the urethra–bladder and the rectum[48].

More recently, the Williams technique[49] of creating a new vagina is indicated for acquired vaginal atresia due to radiotherapy. However, the new vagina is not favourable for coitus.

We have been using McIndoe's technique since 1955. According to some authors, this operation is indicated immediately when there is haematometra and in patients who have attempted sexual intercourse[12,20,25,48,50,51].

Generally speaking, there are two cases in which this anomaly must be corrected – when marriage is imminent and, depending on the case, when the patient is over 17 years old. The reason is to enable the patient to develop her personality fully.

In addition to the use of skin grafts for the new vagina, pelvic peritoneum has also been used[52]. However, it requires opening the abdomen. It is known as Davidov's colpoperitoneoplasty[53]. Sixty-five cases were published by him (in Rothman[54]). Colpoperitoneoplasty had already been used by Kroemer and Stoeckel (in Pimenta Filho[24], Machado[55]), and lately has been used by Pimenta Filho[24].

As mentioned, we prefer and have used McIndoe's operation since 1955. Our present series consists of 110 cases that were operated on. Our surgery has always been done with the cooperation of a plastic surgeon. The skin is obtained from the hypogastrium or the thighs.

#### Use of McIndoe's operation

The operation has two fundamental steps: the preparation of the graft, followed by preparation of the cavity between the rectum and the urethra and bladder[56].

#### Obtaining the graft (plastic surgeon)

Using a dermatome, the plastic surgeon takes a strip of skin large enough to cover the cylindrical mould. The resulting wound on the abdomen or thigh is protected with gauze and cotton and bandaged.

It is preferable to take two grafts from the thighs, one from each side while the patient is in the gynaecological position. A good area of dyschromic skin is obtained which in many cases has less hair than that from the abdomen. In addition, with the patient in the gynaecological position it is preferable for the gynaecologist to prepare first the vaginal space. The plastic surgeon then takes the skin from the thighs. Following this, the skin is placed on the mould and introduced into the prepared cavity. It is better to use an orthoplastic cylinder that weighs 30 g rather than an acrylic one of 300 g.

Another of our plastic surgeons (Professor Orlando Lodovici) prefers to use abdominal skin which is taken with the patient in the semigynaecological position. After the graft is taken the patient is placed in the gynaecological position. The plastic surgeon then covers the mould with the epidermis by suturing the edges with nylon thread.

### The vaginal stage

As we have stated, the patient is placed in the gynaecological position from the beginning. It is preferable for the gynaecologist to first prepare the cavity and then the plastic surgeon takes the graft, prepares the mould, and positions it.

The incision is made with the scalpel transverse to the vestibule or to the retrohymenal fossette, between the anus and the urethral meatus (Figures 10.8 and 10.9). One should first try to cut the bundles of the 'fibrous perineal nucleus' transversely. These are found between the urethra and the

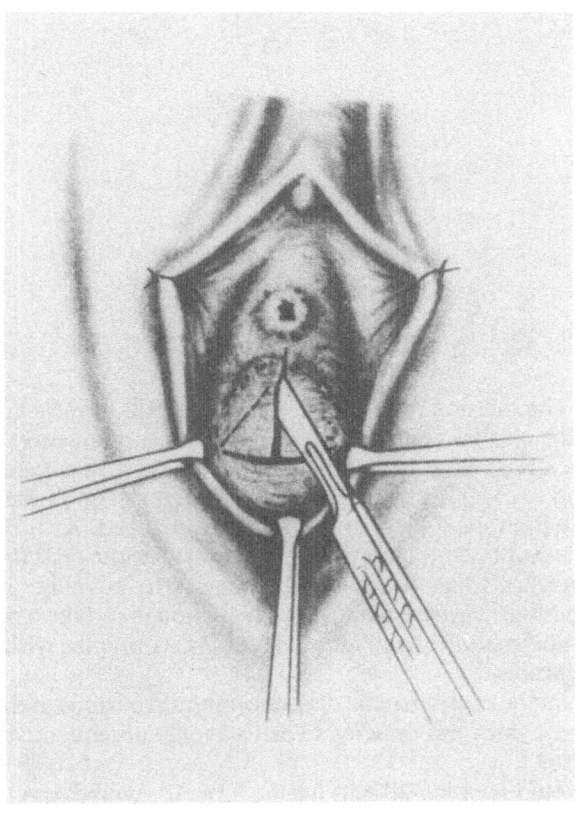

**Figure 10.8** McIndoe's vaginoplasty. Vestibular incision

191

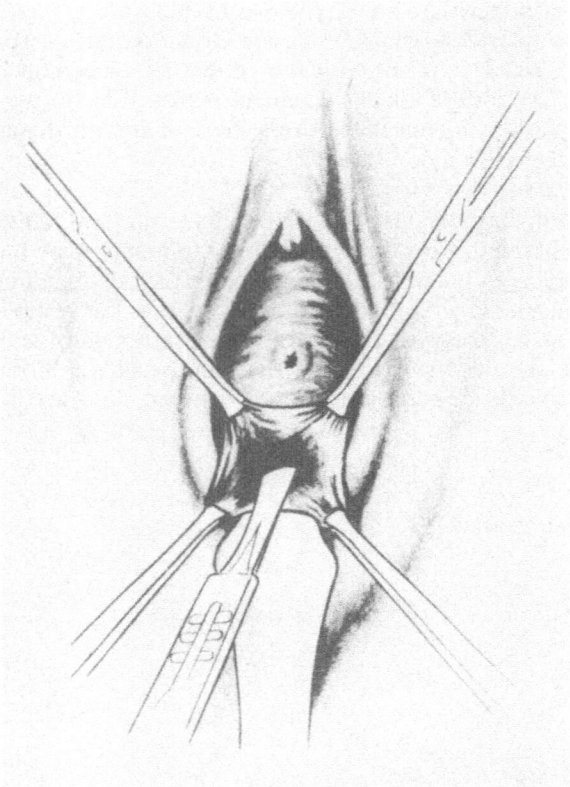

**Figure 10.9**  McIndoe's vaginoplasty. First dissection between anus and urethra

anus/rectum. The incision is deepened and two small tunnels are formed on either side of the urethra (Figure 10.10). These are subsequently united in the midline. By means of the scalpel or its handle, or else with curved-tip scissors, the urethra and bladder are separated from the rectum up to the point where the base of the peritoneal bladder–rectal sac is reached. At this stage of the procedure we never open the urethra, the bladder, or the rectum, an accident which often occurs when the surgeon attempts to advance the dissection through the midline. The cavity must be large enough to take a medium-sized mould. Absolute haemostasis of all small vessels is obtained with plain catgut number 00 ligatures.

Depending on the case, a partial dissection should be done at the sides of the columns of the levator ani muscles so as to avoid subsequent stenosis of the vaginal introitus.

After the mould is prepared and haemostasis is assured and the space has been soaked with physiological saline, the skin-covered mould is introduced (Figures 10.11 and 10.12). The mould is fixed with cloth bandages passing

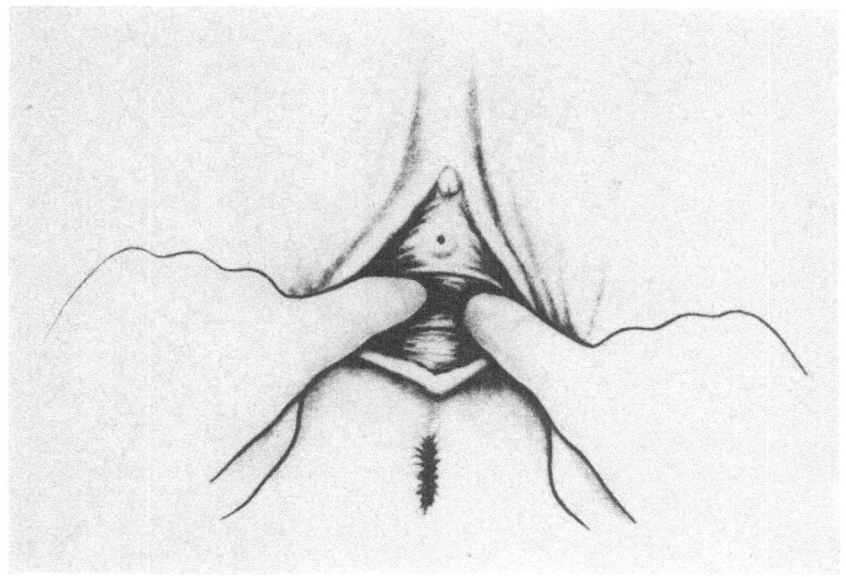

**Figure 10.10** Preparation of the two small tunnels between urethra and rectum

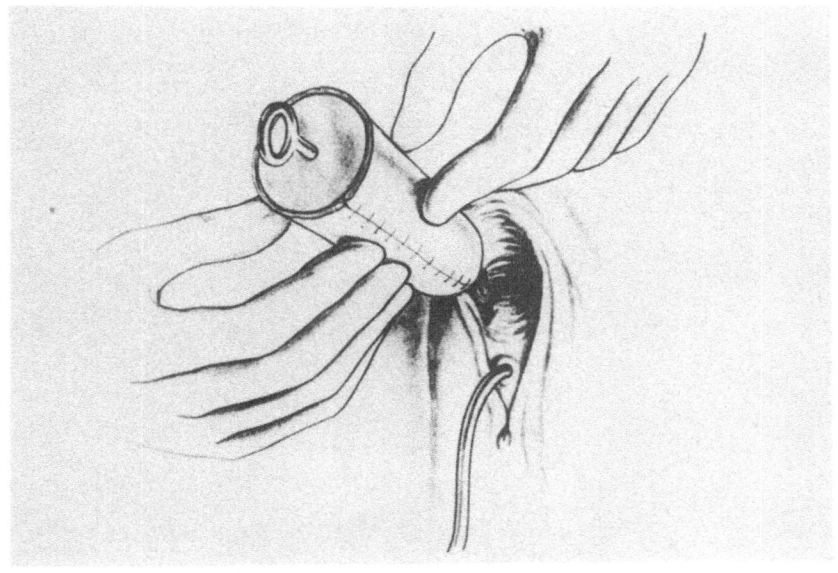

**Figure 10.11** Introduction of the mould covered by skin

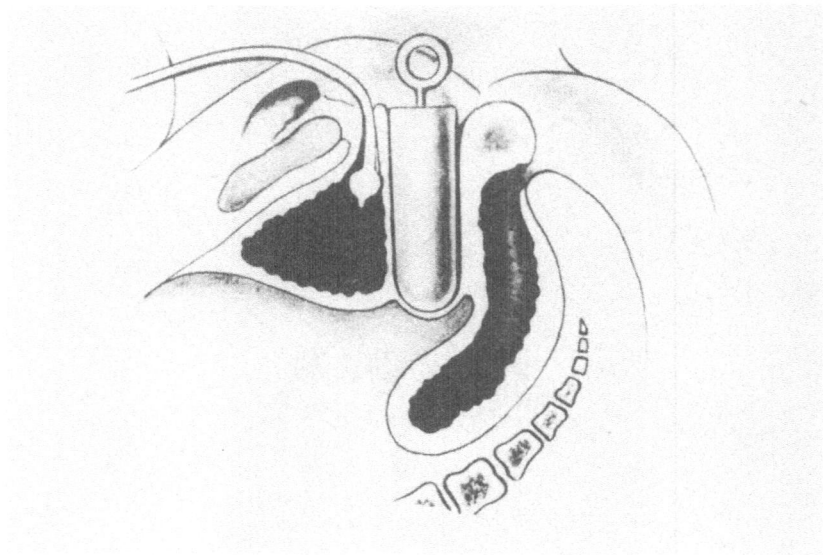

**Figure 10.12** Introduction of the mould covered by skin

through its screw and is attached to the abdominal bandages with adhesive tape (Figure 10.13).

The patient must remain in bed until the 7th or 8th day, after which she is

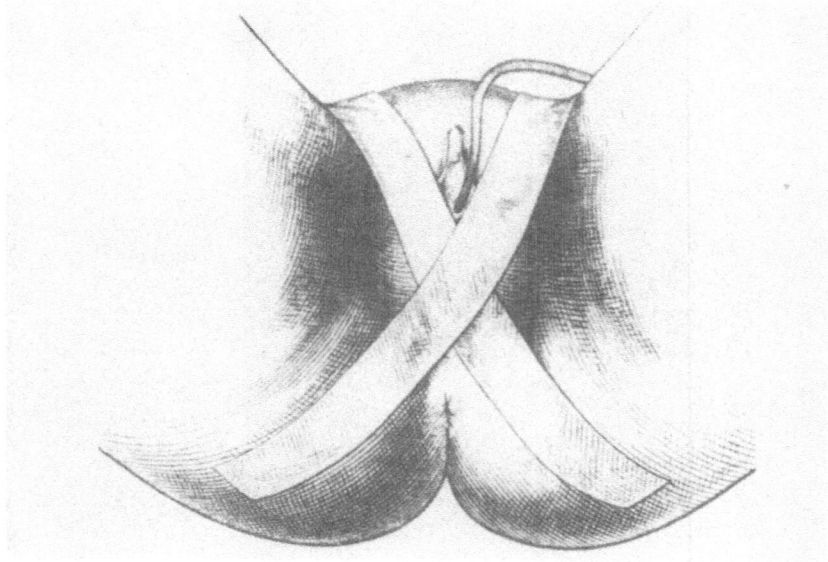

**Figure 10.13** Bandage to fix the vaginal mould

allowed to walk and take out the mould. She must subsequently undergo re-dressing every 48 h. The mould is removed and reinserted. She must be instructed concerning the long term hygiene of the neovagina in order to avoid stenosis.

The acrylic mould is inserted soon after surgery and it remains in place for 8 d. The dressing is changed every other day; after discharge, about the 14th day, the patient herself will prepare and apply the dressings. These consist of removing and reinserting the mould into the neovagina over a period of 6 months.

### Results

The results of the McIndoe's technique are good. In our cases graft integration occurred in 83.3 % of the cases. However, if we add six cases where integration was partial with reoperation necessary, we achieved 90 % of optimum graft integration and optimum condition of permeability. Page and Owley[57] claim 81 % good results.

A late follow-up of four cases, between 1 and 6 years postoperatively, showed 91.3 % satisfactory results in 70 cases. Fifty-four patients had normal sexual relations; 2 had become pregnant and undergone caesarean section. Only 6 cases showed a small vaginal stenosis and 4 revealed dyspareunia. We did not find enterocoele.

Finally, we can say that in our experience the McIndoe's technique is the one offering the best results and the long term use of an acrylic mould guarantees the late success of the operation.

Whitace and Wang[58] reported cyclic cytologic changes in the new vagina and bacterial flora similar to a normal vagina. Ayre[11] also described cyclic ovarian changes; however, Ulfelder[59] did not notice very marked changes. We did not observe cyclic changes in 25 cases. Cytology of the new vagina revealed 18 cases with acidophilic cells between 5 and 20 % and 7 cases between 21 and 40 %. Cells indicative of the luteal phase were not found. Biopsy of the new vagina only showed atrophy of the epidermis (Figure 10.14).

## Vaginal septum and double vagina

Just as in cases of agenesis, cases of transversal septum should be opened with a scalpel by means of a transverse incision. The haematocolpos is opened and drained. After rinsing thoroughly with saline, the fibrous septum is removed followed by circular dissection of the vaginal mucosa between the septum. This is followed by suture of the two margins of the vaginal mucosa to each other, i.e. the lower is anastomosed to the upper (Figure 10.15). It is important to insert a furacin gauze tampon during the first 48 h.

In cases of longitudinal partial or total septum (double vagina), the septum should be removed with a scalpel by means of two longitudinal incisions one on each side of the septum on the anterior and posterior walls.

After the septum is removed, the vaginal mucosa should be dissected from both the anterior and posterior implantation of the septum, and then proceed to the suturing of the margins of the vaginal mucosa to each other. Interrupted sutures of plain catgut 00 should be used. It is also important to leave in place a vaginal tampon of furacin gauze for 48 h.

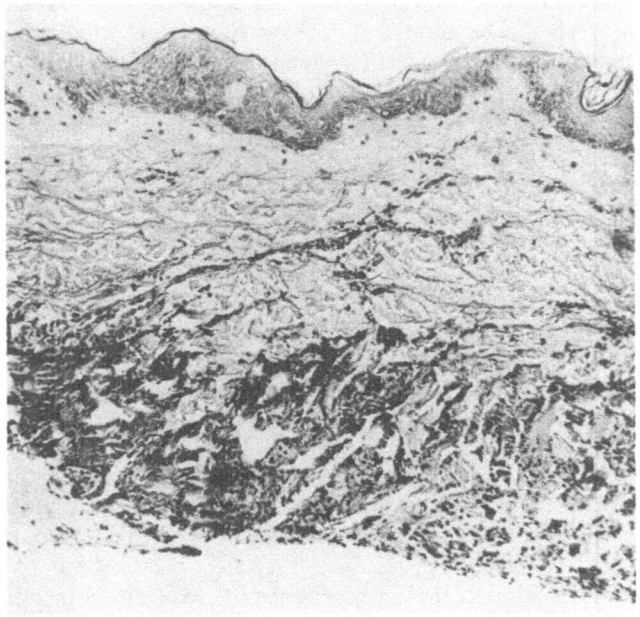

Figure 10.14  Biopsy in neovagina 2 years after the vaginoplasty

The results are usually good. Rarely have vaginal stenosis and dyspareunia been observed.

## Vaginal wolffian cysts

Vaginal wolffian cyst or Gartner's cyst is managed surgically. It should be dissected with a scalpel, care being taken not to perforate it. However, since its

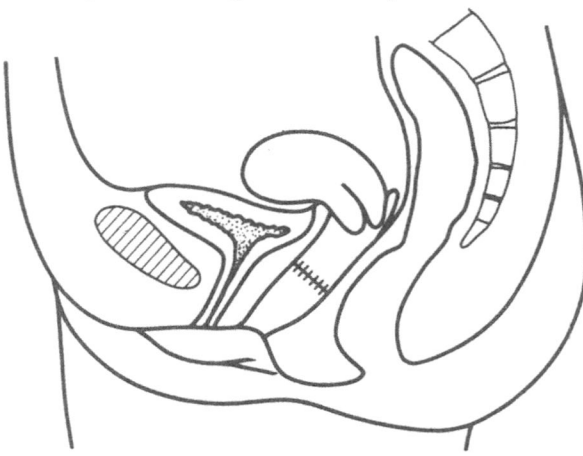

Figure 10.15  Anastomosis between lower and upper vagina in a case of transversal septum

walls are thin it breaks easily. Under these circumstances, the cyst should be stuffed with cotton and closed; this will facilitate the dissection. Some are extremely difficult to dissect because they are elongated and may reach the parametrium laterally. In this case, care must be taken not to injure important vessels and the ureter. Souen and Schivartche[36] observed 43 cases on our service and reported that management by resection of the cyst yielded good results without complications. Only one case which histologically showed adenosis presented recurrence after some time. However, this was satisfactorily resolved by a new surgical resection.

## References

1 Kempermann, C. T. (1935). Beiträge zur Entwicklung des Genitaltraktus der Saüger. *Gegenbaurs Morphol. Jahrbuch.*, **75**, 151
2 Koff, A. K. (1953). Development of the vagina in the human fetus. *Contrib. Embryol.*, **24**, 59
3 Vilas, E. (1932). Über die Entwicklung der menschlichen Scheide. *Z. Anat. Entwick. Ges.*, **98**, 263
4 Nichols, D. H. and Randall, G. L. (1977). *Vaginal Surgery*. (Baltimore: Williams & Wilkins)
5 Azdury, R. S. and Jones, H. W. Jr. (1966). Cytogenetics finding in patients with congenital absence of the vagina. *Am. J. Obstet. Gynecol.*, **94**, 178
6 Linscke, J. A., Curtis, C. and Lamb, E. (1973). Discordance of vaginal agenesis in monozygotic twins. *Obst. Gynecol.*, **41**, 920
7 Le Duc, B., Van Compenhout, J. and Simard, R. (1968). Congenital absence of the vagina. *Am. J. Obst. Gynecol.*, **100**, 512
8 Jost, A. (1958). Embryonic sexual differentiation. In Jones, H. T. and Scott, W. W. (eds.) *Hermaphroditism, Genital Anomalies and Related Endocrine Disorders*, pp. 15–45. (Baltimore: Williams & Wilkins).
9 Jost, A. (1946–1947). Recherches sur la différentiation sexuelle de l'embrion de lapin. *I. Arch. Anat. Micros. Morphol. Exp.*, **36**, 151.
10 Hauser, G. A., Keller, T. H. and Wenner, R. (1961). Das Rokitansky–Kuster Syndrom. *Gynaecologia*, **154**, 111
11 Ayre, J. E. (1944). Cyclic ovarian changes in artificial vagina mucosa. *Am. J. Obstet. Gynecol.*, **48**, 690
12 Bryan, A. L., Nigro, J. A. and Counseller, V. S. (1949). One hundred cases of congenital absence of the vagina. *Surg. Obstet. Gynecol.*, **88**, 79
13 Cohen, H. J., Klein, M. D. and Laver, M. A. (1957). Cysts of the vagina in the new-born infant. *Am. J. Dis. Child.*, **94**, 322
14 Evans, T. N. (1967). The artificial vagina. *Am. J. Obstet. Gynecol.*, **99**, 944
15 Hafez, E. S. E. (1977). The vagina and human reproduction. *Am. J. Obstet. Gynecol.*, **129**, 574
16 Huffman, J. W. (1971). *Ginecologia en la Infancia y en la Adolescencia*, p. 183. (Spanish edition) (Barcelona: Salvat)
17 Jones, H. W. and Scott, W. W. (1958). *Hermaphroditism, Genital Anomalies and Related Endocrine Disorders*, pp. 327–336. (Baltimore: Williams & Wilkins)
18 Salvatore, C. A., Lodovici, O., Spina, V. and Faure, R. (1964). Tratamento da Ausência da Vagina com Especial Referência à Agenesia Vaginal. *Rev. Paul. Med.*, **65**, 63
19 Salvatore, C. A., Lodovici, O. and Gallucci, J. (1967). Ausência Congênita da Vagina. *An. Bras. Ginecol.*, **64**, 317
20 Counseller, V. S. and Davis, C. E. (1968). Atresia of the vagina. *Obst. Gynecol.*, **32**, 528
21 Lodovici, O. (1966). Neovaginoplastia Metodizada no Tratamento da Ausência Congênita de Vagina. *Thesis Medical Faculty, University of São Paulo*
22 Hafez, E. S. E. and Evans, T. N. (1978). *The Human Vagina*. (New York: North-Holland)
23 Semmens, J. P. (1962). Congenital anomalies of female genital tract. *Obstet. Gynecol.*, **19**, 328
24 Pimenta Filho, R. (1975). Sindrome de Rokitansky–Kuster–Hauser. *Thesis*, University of Belo Horizonte

25 McIndoe, A. H. (1959). Discussion in treatment of congenital absence of vagina with emphasis on long-term results. *Proc. R. Soc. Med.*, **52**, 952

26 Capraro, V. J. and Gallego, M. B. (1976). Vaginal agenesis. *Am. J. Obstet. Gynecol.*, **124**, 98

27 Hauser, G. A. and Girotti, M. (1969). Sindrome de Mayer–Rokitansky–Kuster: nuovi aspetti del quadro sintomatologico. *Riv. Ital. Ginecol.*, **53**, 124

28 Salvatore, C. A., Bastos, A. C. and Ramos, L. O. (1978). In Marcondes, P. and Marcondes, E. (eds.) *Pediatria Basica Ginecologia*, Vol. 2, p. 1824. (São Paulo: Savier)

29 Bastos, A. C. and Salvatore, C. A. (1973). Más-formações do aparelho genital feminino. *Mat. Infant.*, **32**, 47

30 Salvatore, C. A. and Ladovici, O. (1978). Vaginal agenesis. *Acta Obst. Gynecol. Scand.*, **57**, 89

31 Cordier, G. (1961). Aplasie du vagin: un decénnie et plus d'expérience. *Bull. Féd. Soc. Gynécol. Obstet.*, **13**, 130

32 Nahoum, J. C. (1968). Disnesias Gonádicas Femininas. IV – Dermatóglifos. *An. Bras. Ginecol.*, **66**, 455

33 Nahoum, J. C., Henriques, C. A., Castelar, D. and Barcelos, J. M. (1970). Vagina e utero rudimentares. *J. Bras. Ginecol.*, **70**, 107

34 Dougherty, C. M. (1972). Anomalies of the vagina. In Dougherty, C. M. and Spencer, R. (eds.) *Female Sex Anomalies*, p. 244. New York: Harper & Row)

35 Evans, M. D. and Paine, G. C. (1965). Quistos e tumores benignos congênitos. *Clin. Obstet. Ginecol.* (Spanish edn.) 997

36 Souen, J. S. and Schivartche, P. L. (1975). Cisto de Gartner. *J. Bras. Ginecol.*, **79**, 143

37 Schubert, G. (1914). Concerning the formation of a new vagina in the case of congenital vaginal malformation. *Surg. Gynecol. Obstet.*, **19**, 376

38 Kirschner, M. and Wagner, G. A. (1930). *Dtsch. Z. Chir.*, **225**, 242 (Or *Abstr. Zentralhlko Ginekol.*, **54**, 2690)

39 Counseller, V. S. and Flor, F. C. (1957). Congenital absence of the vagina. Further results of treatment and a new technique. *Surg. Clin. N. Am.*, **37**, 1107

40 Baldwin, J. F. (1907). Formation of an artificial vagina by intestinal transplantation. *Am. J. Obstet. Gynecol.*, **56**, 636

41 Baldwin, J. F. (1927). The Baldwin operation for formation of artificial vagina. *Surg. Gynecol. Obstet.*, **45**, 569

42 Frank, R. T. and Geist, S. H. (1927). The formation of an artificial vagina by new plastic technics. *Am. J. Obstet. Gynecol.*, **14**, 712

43 Frank, R. T. (1935). The formation of an artificial vagina without operation. *Am. J. Obstet. Gynecol.*, **35**, 1053

44 Frank, R. T. (1938). The formation of an artificial vagina without operation. *Am. J. Obstet. Gynecol.*, **35**, 1055

45 Wharton, L. R. (1946). Difficulties and accidents encountered in construction of the vagina. *Am. J. Obstet. Gynecol.*, **51**, 866

46 Wharton, L. R. (1947). Congenital malformations associated with developmental defects of the female reproductive organs. *Am. J. Obstet. Gynecol.*, **53**, 37

47 Thompson, J. D., Wharton, L. R. and TeLynde, R. W. (1957). Congenital absence of the vagina. *Am. J. Obstet. Gynecol.*, **74**, 397

48 McIndoe, A. H. (1937). The application of cavity grafting. *Surgery*, **1**, 535

49 Williams, E. A. (1964). Congenital absence of the vagina. *J. Obstet. Gynecol. Br.*, **71**, 511

50 Barrows, D. N. (1936). Kirschner–Wagner operation for construction of artificial vagina. *Am. J. Obstet. Gynecol.*, **31**, 156

51 Barrows, D. N. (1957). Results after construction of artificial vaginas. *Am. J. Obstet. Gynecol.*, **73**, 609

52 Machado, L. M. (1937). Vagina artificial com emprego de peritônio pelviano: uma questão de prioridade. *Rev. Ginecol. Obstet.*, **31**, 500

53 Davidov, S. N. (1969). Colpoiesis from the peritoneum of the utero-rectal space. *Obstet. Gynecol. (Moscow)*, 55

54 Rothman, D. (1972). The use of peritoneum in the construction of vagina. *Obstet. Gynecol.*, **40**, 835

55 Machado, L. M. (1936). Nova técnica para formação de uma vagina artificial. *Rev. Ginecol. Obstet.*, **30**, 781

56 Salvatore, C. A. (1974). *Ginecologia Operatria*. (Guanabara: Koogan)

57 Page E. W. and Osley, J. Q. (1969). Surgical correction of vaginal agenesis. *Am. J. Obstet. Gynecol.*, **105**, 774
58 Whitace, F. E. and Wang, Y. Y. (1944). Biological changes in squamous epithelium transplanted to the pelvic connective tissue. *Surg. Gynecol. Obstet.*, **79**, 192
59 Ulfelder, H. (1968). Agenesia of the vagina. *Am. J. Obstet. Gynecol.*, **100**, 745

# 11
# Ambiguous genitalia in the newborn

P. K. KOTTMEIER

The complex pathogenicity leading to intersex problems in infants, often with ambiguous genitalia, reflects the intricate chromosomal, gonadal and hormonal interchanges which lead to the sexual development of the normal infant. The understanding of normal sexual development is essential to permit the identification of the various disorders of sexual development which should be made promptly at birth. Early recognition of abnormal sexual development can be lifesaving in infants with metabolic disorders, such as the salt wasting adrenogenital syndrome, which may lead to death unless treated promptly. The early, correct and definitive gender assignment is imperative since a delay or mistaken assignment may lead to permanent psychological damage to child and family, and may preclude the appropriate gender assignment if the diagnosis has been made too late[1].

## NORMAL SEXUAL DEVELOPMENT

Until recently it was felt that chromosomal disorders played a relatively minor role in the abnormal development of sexual differentiation with the exception of disorders such as Turner's XO syndrome, XX/XY hermaphroditic chimera and other mosaicisms, usually involving XO components. The finding of a 46,XX chromosome complement in a phenotype male with testes, or the presence of testicular tissue in infants with 46,XX true hermaphroditic chromosomal pattern, led to the assumption that gonadal and end organ anomalies were not necessarily related to or caused by chromosomal abnormalities. 'Sex reversal' was felt to be the explanation for the development of a phenotype male with testes, whose chromosomes displayed a 46,XX complement without any evidence of a Y male arm. It was only after Eichwald and Silmser[2] described the HY antigen, which was present in the 'sex reversal' 46,XX male and the true hermaphrodite 46,XX, indicating that in these patients at one time the Y chromosome was indeed present. The short arm of the Y chromosome is thought to be responsible for tubular differentiation, HY antigen formation and male determination. The long arm of the Y is felt to be important in the differentiation of germ cells past the spermatogonia state[3].

The chromosomal pattern of the human, 46,XX chromosomes representing the normal female and 46,XY the normal male, forms the chromosomal basis for various classifications of children with ambiguous genitalia. An infant with ambiguous or abnormal external genitalia with an XX chromosome will be considered to be a female pseudohermaphrodite based on the karyotype. On the other hand, a child with an XY chromosome complement will fall into the male pseudohermaphrodite group. To understand the disorders of sexual differentiation and the interrelationship of chromosomal, gonadal development and hormonal interplay, the gonadal development especially has to be considered.

## GONADAL DEVELOPMENT

The fetal indifferent gonad, under chromosomal influence, will begin to differentiate into ovary or testis at approximately 5–6 weeks of intrauterine age. The primordial germ cells in the male fetus are found in the testicular cord at 6 weeks of intrauterine age and Sertoli cells are found at 9 weeks. The Y chromosome is responsible for testicular formation and abnormalities of the HY function can lead to testicular anomalies, therefore involving either the Müllerian inhibiting substance or testosterone. In the female ovarian stroma and follicles develop between 12 and 16 weeks of intrauterine life. Under the influence of the Müllerian reducing factor[4], the Müllerian anlage will disappear in the male fetus. The second hormone secreted by the fetal testis is testosterone, which stimulates the wolffian duct structures which will eventually develop into the epididymis, the vas deferens and the seminal vesicles. Siiteri[5] has shown that the development of the external male genitalia depends on the conversion of testosterone to dihydrotestosterone through a 5 α-reductase. This conversion of testosterone to the more potent androgen dihydrotestosterone occurs only at the site of the external, but not the internal genitalia, until a later point of fetal development. The internal genitalia therefore depend more on the proximity to the testis to obtain the necessary concentration of testosterone to lead to sexual differentiation. This explains why exogenously administered androgens in the pregnant woman will lead to a masculinization of the external but not the internal female organs, as seen in the female pseudohermaphrodite with adrenogenital syndrome. When the fetal testis lacks the Müllerian reducing factor, the Müllerian anlage will persist, resulting in anomalies such as male pseudohermaphrodites with persistent uterus, dysgenetic testis and external male genitalia. In the course of the normal male development, the external anlage develops under the influence of testosterone into scrotum, penis and glans. In the absence of testosterone, the external anlage will develop into labia majora and minora and clitoris. It was thought that the lower two thirds of the vagina represent the residual portion of the separate urogenital sinus. More recent studies, however, have indicated that at least the upper half of the vagina is Müllerian in origin[6]. However, it should be stressed that not all anomalies of anatomical urogenital development are the result of either chromosomal, gonadal or hormonal anomalies. These unrelated anomalies include the frequently seen labial fusions, imperforate hymen and low vaginal atresias which may lead to hydrocolpos or haematocolpos, and the common vaginal anomalies, including septate vaginas or bifid uteri, in children with imperforate or ectopic ani. Anomalies including uterus and upper vagina, without associated gonadal or chromosomal anomalies, are often the result of a faulty development of the urogenital sinus. While these anomalies can occur separately from the 'abnormal sexual differentiation or development of ambiguous genitalia', they are also frequently present in children with ambiguous genitalia and a distinction between the two main groups, while difficult, is important. The urogenital sinus in the male will form the prostate, the genital tubercle will form the glans penis, the genital folds the urethra and the shaft of the penis, while the genital swellings will form the scrotum. In the absence of androgen, the urogenital sinus leads to the formation of the lower vagina, the genital tubercle to the formation of the clitoris, the genital folds form the labia minora, and the genital swelling results in the labia majora.

## CLASSIFICATION OF ABNORMAL SEXUAL DIFFERENTIATION

### Chromosomal classification

Since chromosomes are responsible for the development of fetal gonads into either ovaries or testes, abnormal chromosomal development can be the cause of faulty gonadal development. While the direct relationship between chromosomal anomalies and Turner's XO syndrome, chimera and patients with mosaicism has long been accepted, it has only recently been established that there is a direct relationship between Y anomalies and testicular development in gonadal differentiation[3,6]. Based on the chromosomal analysis, infants with abnormal genitalia and a set of XX chromosomes with positive sex chromatin (Barr body) are most likely to be female pseudohermaphrodites, or children with true hermaphroditism. A child with a Klinefelter syndrome, which could also fall within this group, does not present with ambiguous genitalia at birth, since it will develop phenotypically as a male.

A child with XY chromosomes and negative sex chromatin is most likely to be a male pseudohermaphrodite, this category including children with either defective androgen production or androgen reaction, or children with defective Müllerian regression. This group also includes the occasional true hermaphrodite or the patient with mixed gonadal dysgenesis. Half of all patients with XO Turner's syndromes also have a negative sex chromatin.

### Gonadal classification

Based on gonadal identification, the following categories of intersex patients occur.

(1) Ovary: Female pseudohermaphrodite
(2) Ovary and testis: True hermaphrodite

(3) Testis: Male pseudohermaphrodite
(4) Testis plus streak: Mixed gonadal dysgenesis
(5) Streak only: Pure gonadal dysgenesis and Turner's syndrome.

## Hormonal classification

Allen[7] classified sexual disorders based on blocks in the synthesis of adrenal, testicular and androgenic steroids, as related to specific enzymatic disorders. These disorders are due either to increased androgenization in the female or to inadequate androgen exposure, function or reaction in the male. Combining the enzymatic deficiencies and the gonads present, Allen proposed the classification shown in Table 11.1.

Walsh and Scott[8] proposed a simplified classification, combining the status of gonads, internal ducts, external genitalia and sex chromatin. The classification includes the following main groups: (1) disorders of gonadal differentiation including Klinefelter syndrome, Turner's syndrome, true hermaphroditism and mixed gonadal dysgenesis, (2) female pseudohermaphroditism, including congenital adrenal hyperplasia, prenatal exogenous virilization from drugs, virilizing disorders of the mother and idiopathic causes and

**Table 11.1  Classification of sexual disorders (Allen[7])**

(1) Female pseudohermaphrodite
   (a) Secondary to endogenous androgen
    (i) 21-dehydroxylase deficiency
    (ii) $11\beta$-hydroxylase deficiency
    (iii) $3\beta$-ol-dehydrogenase deficiency
   (b) Secondary to maternal androgen
    (i) Exogenous administered progestogens
    (ii) Endogenous, virilizing tumors

(2) True hermaphrodite

(3) Male pseudohermaphrodite
   (a) Secondary to inadequate androgen production
    (i) $20\alpha$-hydroxylase deficiency
    (ii) $3\beta$-ol-dehydrogenase deficiency
    (iii) $17\alpha$-hydroxylase deficiency
    (iv) 17,20-desmolase deficiency
    (v) 17-ketosteroid reductase deficiency
   (b) Secondary to inadequate androgen utilization
    (i) Incomplete testicular feminization (Reifenstein, Gilbert-Dreyfuss, Lubs syndromes)
    (ii) Complete testicular feminization syndrome
   (c) Secondary to inadequate conversion of testosterone to dihydrotestosterone. (Pseudovaginal perineoscrotal hypospadias)
   (d) Secondary to deficient MRF (hernia uteri inguinalis)
   (e) Secondary to dysgenetic testes

(4) Mixed gonadal dysgenesis

(5) Pure gonadal dysgenesis
   (a) Turner's syndrome
   (b) XX type
   (c) XY type

(6) Miscellaneous

(3) male pseudohermaphroditism – defective androgen production and defective androgen action including testicular feminization, incomplete testicular feminization, incomplete male pseudohermaphroditism, type 1 and type 2. Defective Müllerian regression with persistent Müllerian duct syndrome also falls within this group (Table 11.2).

## DIAGNOSTIC WORK-UP AND HISTORY

In various forms of sexual disorders, a positive family history can be found, as in patients with adrenogenital syndrome, hermaphroditism or incomplete testicular feminization such as the Reifenstein syndrome. A detailed pedigree analysis is therefore of importance and should be obtained in all instances of ambiguous genitalia. The intake of any drug and signs and symptoms of maternal virilization should be assessed. A positive history of maternal ingestion of androgens during pregnancy can lead to the prompt diagnosis of female pseudohermaphroditism and the appropriate treatment, avoiding pain and anguish for the parents and unnecessary complicated work-up for the infant.

### Physical examination

The physical examination must include the assessment of the total appearance of the infant, not simply the status of the external genitalia and the presence or absence of other associated genital malformations. Examination of the external genitalia should include the evaluation of the size and shape of either clitoris or phallus: even a small phallus usually has only a single frenulum, whereas even the large clitoris usually has a double frenulum. This finding is not helpful, however, in children with hypospadias and the chordee, where the shape of the frenulum may vary. The presence of other associated somatic anomalies, such as webbing of the neck, may indicate syndrome such as Turner's. Increased pigmentation, involving primarily the areola and scrotum, usually suggests an underlying adrenogenital syndrome due to increased ACTH secretion. The shape and size of either labia or scrotum, the presence of a bifid scrotum, and the location of the urethral opening and the presence or absence of a vaginal introitus are also of importance. In a female phenotype, however, the absence of a vaginal opening in itself does not indicate a disorder of sexual development. The presence of a milky vaginal secretion indicates the presence of an infantile uterus. In the neonate, under the influence of maternal oestrogen, the uterus is relatively large and firm and can therefore be palpated with ease on rectal examination. In children with associated major urogenital anomalies, or imperforate or ectopic anus, the palpation of the uterus may not always be possible, however. Sonography does allow the identification of the uterus in almost all instances, whereas the sonographic evidence of normal ovarian development is not reliable. The presence of gonads in the scrotum or low in the labia-scrotal folds usually indicates that the gonads are testes, since ovaries, while easily prolapsing through the internal and external ring, rarely descend into either labia or scrotum. Bilaterally absent testes, or bilateral or unilateral cryptorchidism

Table 11.2 Classification of intersexuality (Walsh and Scott[8])

| Disorder | Gonads | Internal ducts | External genitalia | Sex chromatin |
|---|---|---|---|---|
| *Disorders of gonadal differentiation* | | | | |
| Klinefelter's syndrome | Bilateral testes | Wolffian | Male | Positive |
| Turner's syndrome | Bilateral streaks | Müllerian | Female | 50% neg. |
| True hermaphroditism | Testis + ovary | Mixed | Variable | 80% pos. |
| Mixed gonadal dysgenesis | Testis + streak | Mixed | Variable | Negative |
| *Female pseudohermaphroditism* | | | | |
| Congenital adrenal hyperplasia | Bilateral ovaries | Müllerian | Variable virilization | Positive |
| Prenatal virilization from drugs | | | | |
| Virilizing disorder of mother | | | | |
| Idiopathic | | | | |
| *Male pseudohermaphroditism* | | | | |
| Defective androgen production | Bilateral testes | Wolffian | Variable virilization | Negative |
| 3β-hydroxysteroid dehydrogenase deficiency | | | | |
| 17-hydroxylase deficiency | | | | |
| 17-ketosteroid reductase deficiency | | | | |
| 20,21-desmolase deficiency | | | | |
| 17,20-desmolase deficiency | | | | |
| Defective androgen action | | | | |
| Testicular feminization | Bilateral testes | Absent | Female | Negative |
| Incomplete testicular feminization | Bilateral testes | Wolffian | Variable virilization | Negative |
| Incomplete male pseudo-hermaphroditism, type 1 | Bilateral testes | Partial wolffian | Variable virilization | Negative |
| Lubs syndrome | | | | |
| Gilbert–Dreyfuss syndrome | Bilateral testes | Partial wolffian | Variable virilization | Negative |

| | | | | |
|---|---|---|---|---|
| Reifenstein syndrome | Bilateral testes | Wolffian | Variable virilization | Negative |
| Rosewater syndrome | Bilateral testes | Wolffian | Male | Negative |
| Incomplete male pseudo-hermaphroditism, type 2 | Bilateral testes | Wolffian | Variable virilization | Negative |
| Defective müllerian regression | | | | |
| Persistent müllerian duct syndrome | Bilateral testes | Müllerian and wolffian | Male | Negative |

associated with hypospadias, as well as the 'bilateral inguinal hernia' in a girl with palpable ovaries in either side, should arouse suspicion that the child has a defect of sexual differentiation.

Of utmost importance in a newborn with ambiguous genitalia is the immediate determination whether or not the child belongs to the female pseudohermaphrodites with a salt-losing adrenogenital syndrome, which can lead to rapid dehydration, vascular collapse and death.

## Cytogenetic work-up

The most simple and quickest chromosomal examination consists of the identification of the sex chromatin, a clump of chromatins which is located on the nuclear membrane, called the Barr body. The Barr body, representing the second X chromosome, which is important in the ovarian development, is found in more than 20 % of all females but in less than 2 % of all males. Fluorescent microscopy can also help to identify the Y chromosome and therefore the identification of a male chromosome. A complete karyotyping should be obtained with cultures gained from lymphocytes and a variety of tissues including skin and gonads. Although the karyotype is often not compatible with the patient's eventual gender assignment, its analysis contributes to the understanding of the underlying sexual disorder.

## Biochemical studies

17-hydroxycorticoids, 17-ketosteroids, pregnanetriol, 4-tetrahydro-11 deoxy-cortisol, testosterone, follicle stimulating hormone (FCH), luteinizing hormone (LH) form the basis of hormonal work-up. Sodium and potassium levels are especially important in the child suspected of having a salt-losing adrenogenital syndrome with masculinization of the external genitalia. Although diagnosis of salt-losing adrenogenital syndrome in a virilized female infant with hyponatraemia and hyperkalaemia is simple, it is much more difficult in a male infant where 17-hydroxycorticoids are usually lowered and 17-ketosteroids elevated. Since the difference between normal and abnormal steroids in the newborn may be minimal, the determination of urinary pregnanetriol may also be required.

While all these determinations will be helpful, they are time consuming, yet an early clinical diagnosis can be made in most infants who present with ambiguous genitalia at birth, based on the evaluation of the presence or absence of a uterus and the type of gonads in the newborn infant.

## Radiological work-up

In all infants with ambiguous genitalia the presence or absence and the type of internal genital organ should be confirmed. This can be done by either retrograde urethrography or vaginography, or installation of dye into the urogenital sinus, to detect the presence or absence of a cervix and a uterus. The presence of a cervix invariably indicates the presence of a uterus. In patients in whom neither endoscopy nor radiography allows identification of a uterus, sonography may confirm the clinical finding of a uterus. Sonography will not only allow an identification of a uterus, but also usually show the length and the size of the vagina. Endoscopy still may have to be combined with the radiological work-up in a considerable number of patients, not only to demonstrate a urogenital sinus or the presence or absence of a uterus, but also to identify the exact point of entrance of the vagina into the urethra or vice versa.

## Gonadal biopsy

When physical examination, radiological and endoscopic work-up and cytogenetic and hormonal studies do not allow a definitive diagnosis, laparotomy and gonadal biopsy may be indicated. This is especially true in true hermaphrodites. It is rarely necessary, however, to make a decision as to what gonads have to be removed in infancy, even in children with dysgenetic testes when a final decision can be made at a later date. The laparotomy in neonates is usually limited to diagnostic purposes alone.

## CLINICAL CLASSIFICATION OF SEXUAL DISORDERS LEADING TO AMBIGUOUS GENITALIA IN INFANCY

While the classifications based on chromosomal, gonadal and cytogenetic studies will eventually allow a correct diagnosis, it is imperative that the young infant be diagnosed as soon as possible so that the child with the danger of adrenogenital syndrome is identified early before vascular collapse occurs, and the parents of a newborn child with ambiguous genitalia can be reassured that an early identification of the child's sex can be made, saving embarrassment for the parents and later trauma for the child. For this reason, we have employed a clinical classification which will allow the identification of the most common underlying sexual disorders and the appropriate gender assignment in most instances before sophisticated hormonal studies and multiple tissue cultures have been obtained[9].

It should be stressed that this classification is used for the *early tentative* and not the final diagnosis. This classification is based primarily on the results of physical examination, determining the absence or presence of the uterus, the presence of gonads and the presence of hypospadias. Detailed information of cytogenetic, chromosomal, hormonal and neoplastic problems have been concisely presented by Allen[7], Simpson[10], Hendren[11], and Wilkins[12].

As mentioned before, the infantile uterus, under the influence of maternal hormones, is disproportionately large for the infant size and can therefore be palpated with ease. In addition to its size, its milky excretions usually suggest the presence of a uterus. Even in a child in whom the uterus cannot be palpated because of other associated anomalies, such as an absent rectum, sonography can usually be used to identify not only the upper vagina but also the presence or absence of a uterus.

The clinical classification uses the presence of a uterus as a basic guideline, separating infants with ambiguous genitalia into two major groups – one with and one without an identifiable uterus. The third group includes infants with hypospadias.

# Neonates with ambiguous genitalia, uterus (without testes) (Table 11.3)

## Female pseudohermaphroditism

An infant with ambiguous genitalia and a uterus, but without testes, will most likely fall into the group of female pseudohermaphrodites with adrenogenital syndrome. The adrenogenital syndrome constitutes approximately 70 % of all sexual disorders, and approximately half of these infants display symptoms of salt-wasting or hypertension. Female pseudohermaphrodites are genetically females with a 46,XX karyotype, whose abnormal development of external genitalia is due to exposure to excessive androgens during fetal life. The basic biochemical defect is a block interfering with the production of cortisol, leading to an excessive accumulation of androgenic steroids. Most common is a deficiency of 21-dehydroxylase or $11\beta$-hydroxylase. Others include $3\beta$-ol-dehydrogenase, $17\alpha$-hydroxylase, or adrenal lipoid hyperplasia with a reduced $20\alpha$-hydroxylase or 20,22-desmolase. In the latter group the genitalia are infantile but are rarely ambiguous. Salt losing is associated with a $3\beta$-ol-dehydrogenase or $21\alpha$-dehydrogenase deficiency, paroxysmal hypertension with $11\beta$-hydroxylase deficiency.

The physical findings in infants with pseudohermaphroditism include the presence of ambiguous genitalia with a hypertrophic clitoris and a separate perineal urethra. Hyperpigmentation is often present and a normal uterus is identifiable on physical examination, sonogram or vaginogram. A patient with salt-losing syndrome may present with hyponatraemia, hyperkalaemia, dehydration or vascular collapse. The electrocardiogram in untreated patients may show elevated T-waves and spread QRS complexes. Prompt administration of isotonic saline, glucose, mineral corticoids and hydrocortisone is mandatory to stabilize these infants. In infants with endogenous adrenogenital syndrome, even without salt-losing, virilization will continue if untreated, accompanied by premature closure of epiphysis and hirsutism, necessitating cortisone treatment. Sex rearing is female in view of the genetic make-up, gonads and external and internal genitalia.

The absence of a positive sex chromatin, especially in the presence of a HY antigen, can quickly identify the male pseudohermaphrodite with either dysgenetic testes or mixed gonadal dysgenesis.

Operative therapy in these infants consists of clitoral recession[13]. Clitoral amputation, although an easy cosmetic operation, is no longer acceptable as treatment[14]. In patients with exogenous adrenogenital syndrome, spontaneous regression of the clitoris may occur; operative repair can, therefore, occasionally be postponed to observe spontaneous regression. A vaginoplasty, if minor, can be performed at the time of clitoral recession. Major vaginal reconstruction, seldom necessary in females with adrenogenital syndrome, should await menarche.

## Females with urogenital sinus anomalies

Although isolated embryological malformations involving penis, clitoris, or vagina can occur, they are relatively rare. More common are anatomical genital defects associated with major malformations such as exstrophy, cloaca or imperforate anus, which most often involve the urogenital sinus. The external genitalia may be severely malformed but sex identification is usually possible by inspection alone. Internal disarrangement of vagina, urethra or uterus is common, however, and often requires a detailed endoscopic and radiological work-up. The therapy often depends on the underlying or associated defect.

Genetic sex (46,XX) and the presence of internal female genitalia indicate female sex rearing. Early operative correction of the genital system is not indicated – in contrast to infants with female pseudohermaphroditism – but should either follow or be part of the repair of the usually more major associated anomalies. An exception is constituted by infants with vaginal atresia or imperforate hymen with hydrocolpos at birth. An immediate operative repair is obviously indicated in these children.

### Neonates with ambiguous genitalia, uterus and suspected testes or ovotestes

True hermaphroditism is relatively rare and represents only approximately 10% of all sexual disorders. However, among infants with ambiguous genitalia with uterus and palpable external gonads, true hermaphrodites are the ones most commonly found. Irrespective of the chromosomal complement of the true hermaphrodite, gonads with both ovarian follicles and seminiferous tubules are present. The distribution of the gonadal tissue varies and includes various combinations of ovary, testis or ovotestis. The cytogenetic make-up, often a chromosomal mosaicism, may resemble that seen in male pseudohermaphrodites with dysgenetic testis or mixed gonadal dysgenesis. In over half of true hermaphrodites, however, a 46,XX chromosomal complement is present. In the latter form a positive family history can occasionally be obtained. In almost 80% of true hermaphrodites sex chromatin is positive, although HY antigen is also present.

### Male pseudohermaphroditism

Among infants with ambiguous genitalia, uterus and testis, only two forms of male pseudohermaphrodites have to be considered: those with either mixed dysgenetic testes or mixed gonadal dysgenesis. As male pseudohermaphrodites, they represent individuals with a Y chromosome, and are, therefore, cytogenetically easily distinguishable from female pseudohermaphrodites. Infants with dysgenetic testes often display an XO/XY karyotype, whereas male pseudohermaphrodites with mixed gonadal dysgenesis may have either an XO/XY or 45X/46XY complement[15,16]. Male pseudohermaphrodites without persistent müllerian structures rarely display XO chromosomal complements.

Mixed gonadal dysgenesis is characterized by the presence of a testis and a gonad with a 'streak', similar to the gonadal streak in Turner's syndrome. The chromosomal pattern usually indicates an XO/XY mosaicism but other mosaic patterns have also been reported. Patients with gonadal dysgenesis, which occurs primarily in black children, are usually short and often resemble patients with Turner's syndrome.

The sex assignment in either group is based on the external genitalia. The

**Table 11.3  Neonates with ambiguous genitalia and uterus**

| | Karyotype | Gonads | Sex chromatin | H-Y antigen |
|---|---|---|---|---|
| *With ovaries* | | | | |
| Female pseudohermaphrodite (endogenous or exogenous adrenogenital syndrome) | 46XX | Ovary | + | − |
| 3β-ol-dehydrogenase deficiency; salt loser | | | | |
| 21-dehydroxylase deficiency; salt loser | | | | |
| 11β-hydroxylase deficiency; hypertension | | | | |
| Urogenital sinus anomalies | 46XX | Ovary | + | − |
| Often associated with imperforate anus, exstrophy, and cloaca | | | | |
| *With testes, ovotestis or streaks* | | | | |
| True hermaphrodite | 46XX/46XY or others | Ovary and testis or ovotestis | + | − |
| Male pseudohermaphrodite | | | | |
| Dysgenetic testis | XO/XX | Testes | − | + |
| Mixed gonadal dysgenesis | XO/XX 45XY/46XY | Testes and/or streaks | − | + |

presence of testes may lead to relative masculinization at puberty, although testes tend to degenerate after puberty. The incidence of malignant tumours in dysgenetic testes is extremely high[17]. Therefore a removal of the dysgenetic testis is indicated to avoid later malignancies.

## Neonates with ambiguous genitalia without demonstrable uterus (Table 11.4)

### Male pseudohermaphroditism with lipoid hyperplasia

While the female pseudohermaphrodite with lipoid hyperplasia does not display sexual ambiguity, all males reported have either external female genitalia with a small clitoris with a short blind ending vagina or ambiguous genitalia[7]. Testes are often undescended; the skin usually shows increased pigmentation. The hormonal block is assumed to be a lack of conversion of cholesterol to pregnenolone with a reduction of all steroids. A severe salt loss is found in all of these patients and only a few long term survivors are known[11]. The diagnosis should be suspected in an infant with moderate anomalies of the external genitalia with increased skin pigmentation and salt losing syndrome; urinary 17-ketosteroids and 17-hydroxycorticoids are decreased, and adrenal cholesterol increased.

### Incomplete testicular feminization (Lubs, Gilbert–Dreyfuss, Reifenstein syndromes)[18-21]

Testicular feminization occurs in children with ambiguous genitalia who represent male hermaphrodites with inadequate virilization and development of gynaecomastia at puberty, similar to that seen in the complete testicular feminization syndrome. In contrast to the latter, where the external genitalia appear to be normal female, there is often a family history, compatible with an X-linked recessive trait, of relatives with congenital sexual ambiguity and gynaecomastia. In the postpubertal patient an elevated LH is present in the face of a normal or elevated serum testosterone. In the younger infant more specific tests for androgen insensitivity, the basic problem, are required. Most infants in this group will not be diagnosed until puberty, and only in children in whom the sexual ambiguity is more pronounced and where the family history will give a clue to the underlying pathology, will the various syndromes be suspected.

Table 11.4  Neonates with ambiguous genitalia without uterus

| Male pseudohermaphrodite | Karyotype | Gonads | Sex chromatin | H-Y antigen |
|---|---|---|---|---|
| Lipoid adrenal hyperplasia | | | | |
| 20α-hydroxylase deficiency; salt loser | XY | Testes | − | + |
| Incomplete testicular feminization | | | | |
| (Lubs, Gilbert–Dreyfuss syndromes) | XY | Testes | − | + |
| Dysgenetic testes | | | | |
| Without MRF deficiency | XY | Testes | − | + |

213

## Hypospadias (Table 11.5)

Depending on the size and type of hypospadias, the newborn with hypospadias may have the appearance of a normal male on one end of the spectrum, or that of a female on the other end. The majority of neonates with hypospadias, a failure of distal migration of the urethra, have a normal sexual differentiation if both testes are descended. The defect represents only a localized anatomical defect of the ventral portion of the urethra. However, in patients with a urethral opening at the scrotal junction or perineum and cryptorchidism, especially if bilateral, the association with major disorders of sexual differentiation has been reported in 25–50%[9]. Rajfer[22] reported an incidence of intersexuality of 53% in patients with hypospadias, ambiguous genitalia and cryptorchidism. It included male pseudohermaphroditism, mixed gonadal dysgenesis, true hermaphroditism, female pseudohermaphroditism and gonadal agenesis. In patients without ambiguous genitalia, hypospadias and cryptorchidism, the incidence of sexual disorders was close to 30%. Intersexuality is twice as high in children with hypospadias and bilateral cryptorchidism as in those with unilateral cryptorchidism. Hypospadias as a prominent finding has also been reported, as shown in Table 10.3, associated with numerous disorders. It occurs with many autosomal and dysmorphic syndromes, including facial anomalies and/or hypertelorism.

## OPERATIVE TREATMENT OF NEONATAL GENITAL AMBIGUITY

Laparotomy and/or gonadal biopsy is indicated in only a small number of neonates with sexual ambiguity, such as the true hermaphrodite or the male pseudohermaphrodite with dysgenetic testes or gonadal dysgenesis. In children with adrenogenital syndrome, the physical examination and radiological, hormonal and chromosomal work-up should be adequate to obtain a diagnosis without laparotomy. Sex determination should be made promptly at birth to place the neonate in the most appropriate sex role, usually compatible with his external genitalia. In the absence of a phallus the surgical construction of a functional penis is extremely difficult, if possible at all[10]. If the child's external genitalia do not coincide with the assigned sex, however, the child's external appearance should be corrected promptly to spare parents, relatives and the child possible embarrassment.

Vaginal anomalies, such as vaginal atresia or imperforate hymen leading to complications such as hydrocolpos, will require early operative correction in infancy. In the absence of this complication a complete reconstruction of the vagina can and usually should be postponed until the female is capable of

Table 11.5  Neonates with hypospadias

| |
| --- |
| Male pseudohermaphrodite |
| Female pseudohermaphrodite |
| True hermaphrodite |
| Mixed gonadal dysgenesis |
| Gonadal agenesis |
| Part of somatic syndromes (such as Smith, Opitz, Aarskog, Scott, Robinow) |

keeping the vagina dilated or prior to an expected menarche. The routine cosmetic procedure in females with hypertrophic clitoris or phallus in the past consisted of a complete clitorectomy. While this is cosmetically a simple operative procedure, it is now considered unacceptable[13] since regardless of its shape the clitoris will remain the main or only source of sexual gratification, especially in the absence of a normal vagina.

A recession should retain innervation and erectability as the basis for adequate sexual function. Minor complications or discomfort during erection, such as the appearance of a bulge of the recessed clitoris during erection, are not uncommon. This seems to be a small price to pay, however, when compared with visually satisfactory female genitalia achieved by complete removal of all erectile and innervated tissue and hence without the potential of orgasm.

The early recession in childhood allows a girl to grow up without embarrassment. At a time and age when a patient is capable of evaluating her own sexual response, a further surgical repair, if indicated, can then be performed.

## SUMMARY

Although the neonate with sexual ambiguity may occasionally represent a challenging diagnostic problem, the vast majority can be recognized without delay. Most neonates with external genital anomalies belong to a group of female pseudohermaphrodites in whom simple diagnostic manoeuvres, as described, can lead to a prompt diagnosis. The anticipation of a possible salt-losing syndrome should allow the physician to make the diagnosis and to initiate therapy without having to wait for complicated serum hormonal levels. To paraphrase Hendren's statement, 'There is no need to await the determination of plasma testosterone concentration in a virilized pigmented 9-day-old infant in circulatory collapse, whose electrocardiogram shows peaked T-waves and spread QRS complexes'[10]. The classification used in this report hopefully should enable even the physician with little experience to initiate the most appropriate work-up, leading to an early diagnosis and to a decision as to whether to raise the child as male or female.

While chromosomal studies, biochemical determinations and evaluation of internal gonads are necessary to understand the nature of a sexual disorder, the determination of future sex role of the infant is almost invariably based on the shape of the external genitalia. A definite sex role assignment should be made promptly to spare family and child anguish, pain and embarrassment, keeping in mind that a late sex change, after the age of 1–2 years, is usually disastrous to the child.

## References

1 Money, J., Hampson, J. G. and Hampson, J. L. (1955). Hermaphroditism: Recommendations concerning assignment of sex, change of sex and psychologic management. *Bull. Johns Hopkins Hosp.*, **97**, 284
2 Eichwald, E. and Silmser, C. (1955). Skin. *Transplant. Bull.*, **2**, 148
3 Donahoe, P. M., Crawford, J. D. and Hendron, W. (1977). True hermaphroditism: A clinical

description in a proposed function for the long arm of the Y chromosome. *J. Pediatr. Surg.*, 13, 293

4 Jost, A. (1960). The role of fetal hormones in prenatal development. *Harvey Lect.*, 55, 201

5 Siiteri, P. K. and Wilson, J. D. (1974). Testosterone formation and metabolism during male sexual differentiation in the human embryo. *J. Clin. Endocrinol. Metab.*, 38, 113

6 Donahoe, P. K., Crawford, D. J. and Hendron, H. W. (1979). Mixed gonadal dysgenesis, pathogenesis and management. *J. Pediatr. Surg.*, 14, 287

7 Allen, T. D. (1976). Disorders of sexual differentiation. *Urology,* 7 (Suppl.), 4

8 Walsh, P. C. and Scott, W. W. (1979). *Intersex. Pediatric Surgery.* 3rd edn., Vol. 2, p. 1411. (Chicago: Year Book Medical)

9 Kottmeier, P. K. and Velcek, F. T. (1978). Ambiguous genitalia in the neonate. *Clin. Perinatol.*, 5, 163

10 Simpson, J. L. (1976). *Disorders of Sexual Differentiation.* (New York; Academic Press)

11 Hendren, H. W. and Crawford, J. D. (1972). The child with ambiguous genitalia. *Curr. Probl. Surg.*, November

12 Wilkins, L. (1965). *The Diagnosis and Treatment of Endocrine Disorders in Childhood and Adolescence.* 3rd edn. (Springfield, Ill.: Charles C. Thomas)

13 Randolph, J., and Hung, W. (1970). Reduction clitoroplasty in females with hypertrophied clitoris. *J. Pediatr. Surg.*, 5, 224

14 Shaw, A. (1977). Subcutaneous reduction clitoroplasty. *J. Pediatr. Surg.*, 12, 331

15 Sohval, A. R. (1963). 'Mixed' gonadal dysgenesis, a variety of hermaphroditism. *Am. J. Hum. Genet.*, 15, 155

16 Sohval, A. R. (1963). Hermaphroditism with 'atypical' or 'mixed' gonadal dysgenesis, relationship to gonadal neoplasm. *Am. J. Med.*, 36, 281

17 Scully, R. E. (1970). Gonadoblastoma: A review of 74 cases. *Cancer*, 25, 1340

18 Gilbert-Dreyfuss, S., Sebaoun, A. A. and Belaisch, J. (1957). Étude d'un cas familial d'androgyroidisme avec hypospadias grave, gynécomastie et hyperestrogénie. *Ann. Endocrinol. (Paris)*, 18, 93

19 Lubs, H. A. Jr., Vilar, O. and Bergenstal, D. M. (1959). Familial male pseudohermaphroditism with labial testes and partial feminization, endocrine studies and genetic aspects. *J. Clin. Endocrinol. Metab.*, 19, 1110

20 Reifenstein, E. C. Jr. (1947). Hereditary familial hypospadism. *Proc. Am. Fed. Clin. Res.*, 3, 86

21 Wilson, D., Harrod, M. J., Goldstein, J. L., Hemsell, D. L. and MacDonald, P. C. (1974). Familial incomplete male pseudohermaphroditism, type 1. *N. Engl. J. Med.*, 290, 1097

22 Rajfer, J. Jr. and Walsh, P. C. (1976). The incidence of intersexuality in patients with hypospadias and cryptorchidism. *J. Urol.*, 116, 769

# Index

217

# INDEX

bilirubin
  serum level 125
  unconjugated, serum levels of 93
bladder
  accessory urethra egresses from 156
  epispadias and exstrophy of 145
  exstrophy of 152, 160
  fasciculated 153
  open (lower coelosomia) 132
  sacculated 153
blood flow, pulmonary 5
blood gas
  levels 124
  samples 142
blood group antigens 94
blood groups 94
blood supply
  arterial
    aberrant, to sequestration 57
    anomalies of 56
    from aorta 64
    increase in 58
    variations in 56
  to lesion, abnormalities of 59
  systemic arterial 55, 57
blood vessels
  enlarged 37
  vaginal 180
bowel
  apple peel 136
  atresia of 136
  congested mass of, plus atresia 140
  continuity 138
  dilated loops of 121
  distal to atresia 126
  gross distension of 121
  ischaemia or perforation of proximal
    dilated 123
  length, residual 129
  malrotation 138
  neonatal small 129
  normal, anatomy 110
  oedema of 139
  in omphalocoele 131
  patency of lumen of distal 116
  protruding 132
  recanalization of 109
  short, syndrome 128
  small 126
brachiocephalic (inominate) artery 20
  large (right) 27-8
  origin of 20
brain size, growth in 128
breast, extra 189
brevicollis
  lungs in 40
  syndrome 17
bronchi

compression of 37
connective tissue sheaths around 15
intrasegmental 38
lobar 7, 38
lobular, level of 8
main 18
  and lobar, development of 31
  obstruction of major 81
segmental 8
stenosis of 37
bronchial atresia 37-8, 42
bronchial branch patterns 19, 31-8
bronchial branching patterns
  abnormal segmental and peripheral
    36-7
  intrapulmonary 38
bronchial coarctation 37
bronchial deficiencies, segmental 9
bronchial developmental errors 38
bronchial duplication 58
bronchial stenoses 37
bronchial tree
  central 7
  generations of 76
bronchiectasis 15, 60
bronchioles 8, 81
bronchioli, terminal 76
bronchiolitis 19
  in infants 15
bronchobiliary fistula, congenital 33
bronchogenic cyst 75-89
  asymptomatic 85
  carinal or hylar 77
  of cervical region 77, 83
  childhood 84
  classification of 75
  clinical aspects of 84-8
    diagnostic studies 86-8
    examination of 85
    treatment of 88
  diagnosis of symptomatic 85
  embryology 76-8
  incidence of 75
  infection of, in infancy 83
  intrapericardial 77
  intrapulmonic 77, 81
  intraspinal 77
  intrathoracic 85
  mediastinal 78, 79
    infection of 84
  morbidity from surgical therapy 88
  multiple 81
  natural history 79-84
  in neck 84
  neonatal 83
  occurrence of 75
  in paediatric age group 85
  paraoesophageal 77, 88

219